to my friend & support

Kind regards

Jenny Mies.

Illustrated Glossary of Petroleum Geochemistry

Illustrated Glossary of Petroleum Geochemistry

Jennifer A. Miles
Research Fellow
University of Reading

CLARENDON PRESS · OXFORD
1989

Oxford University Press, Walton Street, Oxford OX2 6DP
Oxford New York Toronto
Delhi Bombay Calcutta Madras Karachi
Petaling Jaya Singapore Hong Kong Tokyo
Nairobi Dar es Salaam Cape Town
Melbourne Auckland
and associated companies in
Berlin Ibadan

Oxford is a trade mark of Oxford University Press

Published in the United States
by Oxford University Press, New York

British Library Cataloguing in Publication Data
Miles, Jennifer A.
Illustrated glossary of petroleum
geochemistry
1. Petroleum deposits. Geochemical aspects
I. Title
553.2'82
ISBN 0–19–854492–8

Library of Congress Cataloging in Publication Data
Miles, Jennifer A.
Illustrated glossary of petroleum geochemistry
/ Jennifer A. Miles.
p. cm. Bibliography: p.
1. Petroleum—Geology—Dictionaries. 2. Geochemical prospecting—Dictionaries. I. Title.
TN870.5.M493 1989 553.2'82'0321—dc19 88–35712
ISBN 0–19–854492–8

Set by Hope Services, Abingdon
Printed in Great Britain by
Biddles Ltd, Guildford & King's Lynn

Preface

For ten years I worked as a petroleum geochemist in exploration and research departments for major oil companies. During that time my responsibilities were threefold: the application of petroleum geochemistry; the training of geologists, geophysicists, and engineers to understand and apply geochemistry; and the development of new, interpretive techniques for exploration. I learned that there were three main sources of confusion for those on the exploration side: the nomenclature of chemical compounds and the 'chemical' side of geochemistry; the multitude of synonyms for a small number of entities; and the individual approach that each company adopts to the teaching of geochemistry. Most companies have their own preferred nomenclature, so that when confronted with geochemical reports from external sources geologists have to learn a fresh set of terms. The nomenclature of the subject has also changed with time. Early geochemical reports contained mainly terms drawn from the application of coal science to oil exploration, whereas later reports, especially those written since the introduction of gas chromatography–mass spectrometry, are highly chemical. Early reports may, however, be the only source of geochemical data as samples from older wells are often used up, preventing modern types of geochemical analysis.

Only the larger oil companies can afford to have a full-time geochemist to deal with problems of this kind. In medium and small companies the task is left to the geologist, whose resources may be confined to a copy of Tissot and Welte or Hunt and a one-week course in geochemistry.

One problem with the existing literature is the difficulty of obtaining rapid access to information. The enquirer needs to know which of a multitude of entries in an index will provide what is needed. All too often the quest is abandoned unless the information is quickly and obviously to hand.

I have produced this text in the hope of providing ready access to a basic common language so that geoscientists and management may communicate with geochemists and their literature. The glossary will also, I trust, bridge the gap between the non-specialist and the comprehensive textbook. If we can improve communications between the practitioners of petroleum geochemistry and those who use the results of their work, much will have been gained.

I would like to thank the Robertson Group and Exploration Logging for valuable assistance with some of the illustrations for the glossary.

Reading J.A.M.
October 1988

Contents

Structure of the book

The text is divided into two parts. The first is a summary section in which the tables provide a simple overview of some aspect of geochemistry, to provide a quick reference. These tables are not exhaustive and the reader is referred to the list of books and journals for reading and reference.

The main section is the glossary: an alphabetical list. Acronyms and abbreviations are treated as words. Under each entry the term is defined and its relevance to petroleum geochemistry explained, with, where appropriate, an illustration. Synonyms are listed, and any other terms that may supply additional information. An attempt has been made to keep the entries self-contained, but also short. Terms that are explained in the glossary are indicated by asterisks on their first occurrence under each heading. Items for which there are many synonyms are explained under their most common usage. A reference is given at the end of the entry for most of the terms. These are intended to indicate a source of literature that the reader may wish to follow up. Some of the simple chemical entries are merely referred to a textbook of organic or physical chemistry. Examples of such texts are given in the recommended reading list and general references. Where there is a reasonable account of the application or a good summary paper, this has been quoted. Many of the references are to journals that are generally available rather than to original or landmark publications, since many of these are in obscure journals. In a few instances an obscure reference is quoted where it is the only available source. Many useful references will be found in the papers recommended for further investigation.

Summary

Major measured bulk maturation parameters—Oil prone source rocks

	Spore colour index	TAI	VR_0 %	Extract/ TOC‰	Alkane/ extract %	CPI Carbon Preference Index	LOM
Immature	1–3.5	<2.2	<0.5	<70	<25	>1.5	<7.0
Early mature	3.5–5.0	2.2–2.3	0.5–0.65	70–100	25–30	1.5–1.2	7.0–8.0
Peak mature	5.0–7	2.3–2.6	0.65–0.9	>100	30–50	1.2–1.0	8.0–10.0
Late mature	7–8.5	2.6–3.5	0.9–1.3	100–50	50	1.0	10.0–11.5
Post mature	8.5–10	>3.5	>1.3	<50	<50	1.0	>11.5

TAI = Thermal Alteration Index; VR = Vitrinite reflectance;

Major measured maturation parameters—Gas prone source rocks

	Thermal Alteration Index	Vitrinite reflectance VR_0%	H/C ratio (III)
Immature	<2.5	<0.7	>0.9
Early mature	2.5–2.6	0.7–1.3	0.9–0.85
Peak mature	2.6–3.6	1.3–2.2	0.85–0.5
Late mature	3.6–4.0	2.2–3.0	0.5–0.25
Post mature	>4.0	>3.0	<0.25

Major molecular ratios—Oil prone source rocks

Triterpanes					Steranes	
C_{27} T_s/T_m	C_{30} ββ/total	C_{30} Moretane/ hopane	C_{31} 22*R*/22*S* Homo- hopanes	$\frac{C_{29}+C_{30}}{C_{27}}$	C_{27} 20*S*/20*R*+20*S* 13β17α Diacholestane	C_{28} 24*S*/24*S*+24*R* Methyl cholestane
<0.1	40	0.7	<0.2	>5	0.45	0.6
<0.4 Ca ~0.6 Sh	5 coals 0 other	0.5	1.0	3.0	0.55	0.5=m
0.6 Ca 1.0 Sh	–	0.15–0.1	1.5	2.0	0.6=m	
>2.0	–	–	–	1.0		
–	–	–	–	–		

Rock Eval production index %	T_{max}°C		H/C ratio		Temperature °F		
	I	II/III	I	II	slow sed.	fast sed.	
<0.1	<435 ±10	<427 ±10	1.6	1.2	<145	<200	Immature
0.1–0.2	435–444 ±10	427–435 ±10	1.6–1.45	1.21–1.1	145–190	200–250	Early mature
0.2–0.4	444–455 ±10	435–442 ±10	1.45–1.0	1.1–0.9	190–265	250–320	Peak mature
0.4	>455	>442	1.0–0.8	0.9–0.8	265–320	320–375	Late mature
–	–	–	0.8–0.5	0.8–0.5	>320	>375	Post mature

CPI = Carbon Preference Index

T_{max} °C	LOM	Temperature °F	MPI	
<440±15	<9.0	<190	<0.5	Immature
440–460 ±15	9.0–11.5		0.5–1.5	Early mature
460–530 ±15	11.5–14.5	~350	<1.5	Peak mature
>530 ±15	14.5–18.0		–	Late mature
	>18.0	>500	–	Post mature

MPI = Methylphenanthrene Index

Steranes		Aromatic steroids			
C_{29} 20*S*/20*R*+20*S* Ethyl cholestane	C_{29} 20*R*+20*S* αββ/αββ+ααα	C_{28} Tri / C_{29}mono+C_{28}tri	C_{20} Tri / C_{20}tri+C_{28}tri	Vanadyl porphyrins DPEP/Etio	
<5	<0.2	0	<0.2	>7	Immature
20	0.47	0.5	0.2	5	Early mature
40–55	0.6–0.8	0.65	0.35	2	Peak mature
	0.8–1.0	1.0=m	0.6	0	Late mature
			1.0	–	Post mature

Ca Calcareous
Sh Shale
TTI values are not measured maturation parameters.

Key to molecular ratios

Carbon number	Ratio		Definition
C_{30}	ββ/Total	:	17β (*H*), 21β (*H*) C_{30} hopane/Total C_{30} hopanes
C_{30}	Moretane/hopane	:	17β (*H*), 21α (*H*) C_{30} moretane/17α (*H*), 21β (*H*) C_{30} hopane
C_{31}	22*S*/22*R* Homohopanes	:	22*S* 17α (*H*), 21β (*H*) homohopane/22*R* 17α (*H*), 21β (*H*) homohopane
C_{27}	20*S*	:	20*S* 13β (*H*), 17α (*H*) diacholestane
	20*S* + 20*R* Diacholestane		20*S* 13β (*H*), 17α (*H*) + 20*R* 13β (*H*), 17α (*H*) diacholestane
C_{28}	24*S*/24*S* + 24*R* Methyl cholestane	:	24*S*, 24-methyl - 5α (*H*), 14α (*H*), 17α (*H*) 20*R* cholestane
			24*S* + 24*R* 24-methyl - 5α (*H*), 14α (*H*), 17α (*H*) 20*R* cholestane
C_{29}	20*S*/20*S* + 20*R* Ethyl cholestane	:	20*S* 24-ethyl - 5α (*H*), 14α (*H*), 17α (*H*) cholestane
			20*S* + 20*R* 24-ethyl - 5α (*H*), 14α (*H*), 17α (*H*) cholestane
C_{29}	20*R* + 20*S* αββ/αββ + ααα	:	20*R* + 20*S* 24-ethyl 5α (*H*), 14β (*H*), 17β (*H*) cholestane
			20*R* + 20*S* 24-ethyl 5α (*H*), 14β (*H*), 17β (*H*) + 24-ethyl 5α (*H*), 14α (*H*), 17α (*H*) cholestane
C_{28}	Tri	:	C_{28} triaromatic steroids
C_{28}	Tri + C_{29} mono		C_{28} triaromatic steroids + C_{29} monoaromatic steroids
C_{20}	Tri	:	C_{20} triaromatic steroids
C_{20} + C_{28}	Tri		C_{20} triaromatic steroids + C_{28} triaromatic steroids
DPEP/Etio		:	DPEP vanadyl porphyrins/Etio vanadyl porphyrins

Transformation of organic matter

	Protein	Carbohydrates	Lipids	Pigments	Lignin	
Living organic matter		cellulose polysaccarides	waxes fatty acids, esters terpenoids steroids	β-carotane chlorophyll		– Diagenesis ↓ reactions – biochemical (eg. fermentation) biodegradation condensation polymerization
Soils and near-surface sediments	Amino acids	Sugars	Lipids	Pigments	Humin, humic and fulvic acids	
			waxes alkanes terpanes steranes	β-carotane porphyrins pristane phytane		– Catagenesis – chemical reactions ↓ and cracking
Sedimentary rocks		Soluble organic matter oil and gas		Kerogen		
Metamorphic rocks		Carbon/graphite				– Metagenesis chemical reactions

Correlation of kerogen type nomenclature

Sub-aqueous	Algal	Liptinite or Exinite	Alginite (*Botryococcus Tasmanites*)	Algal sapropel	Type I kerogen
	Herbaceous stem		Sporinite Cutinite Suberinite Liptodetrinite Sapropel Bituminite Resinite Amorphinite L (Dinoflagellates) Amorphous	Waxy sapropel	Type II kerogn
Sub-aerial	Woody	Vitrinite	Vitrinite {Telinite, Vitrodetrinite, Collinite} Amorphinite V	Humic	Type III (A) kerogen
	Coaly	Inertinite	Inertinite {Semifusinite, Fusinite, Pyrofusinite, Inertodetrinite, Macrinite, Micrinite}	Humic	Type IV or IIIB kerogen

Exsudatinite, fluorinite, and bitumen are not strictly kerogen

Summary of analytical procedures for oils and source rock extracts

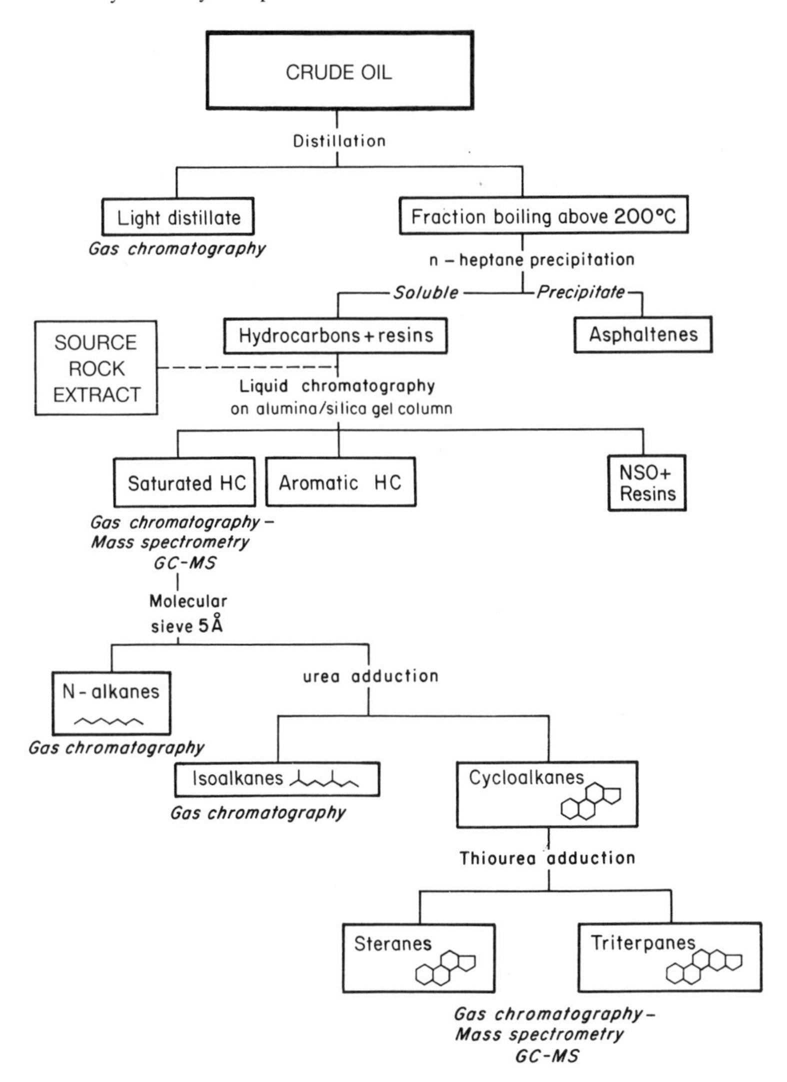

SUMMARY

Source rock characterization

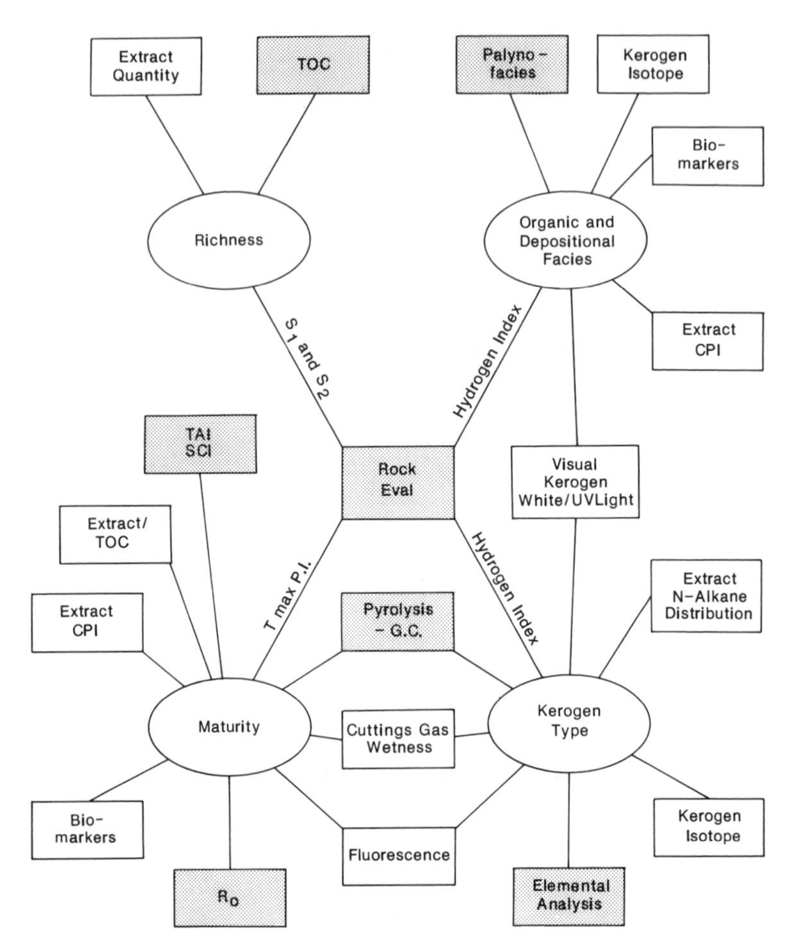

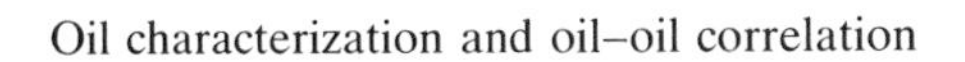

Oil characterization and oil–oil correlation

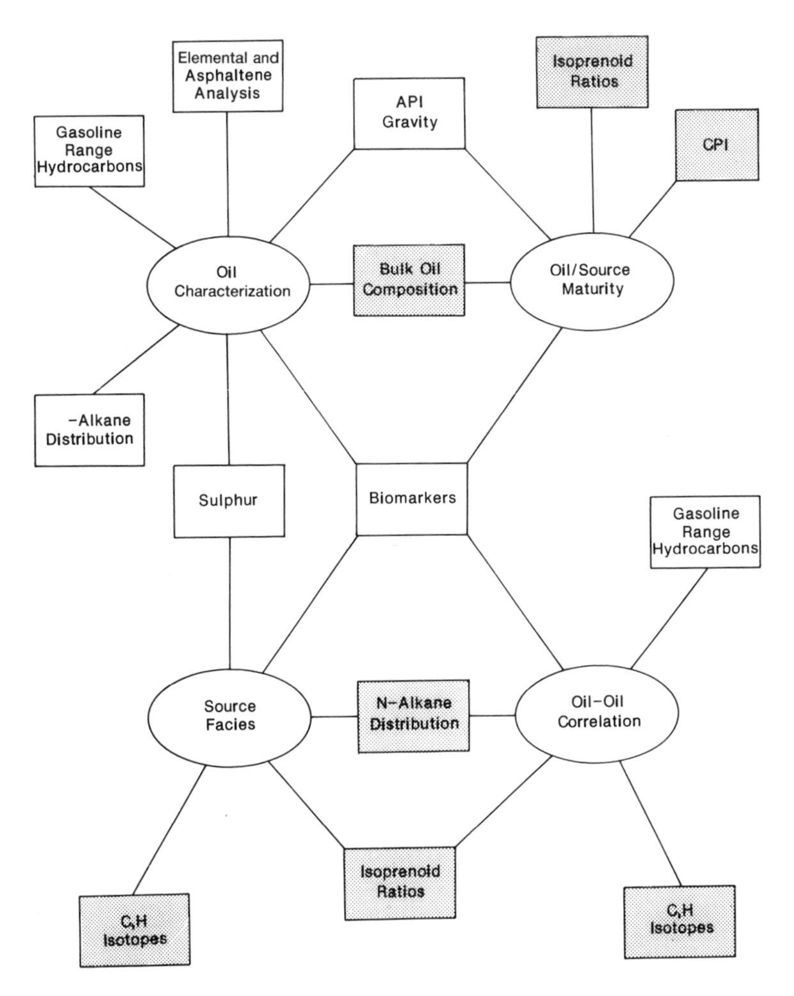

SUMMARY

Oil–source rock correlation

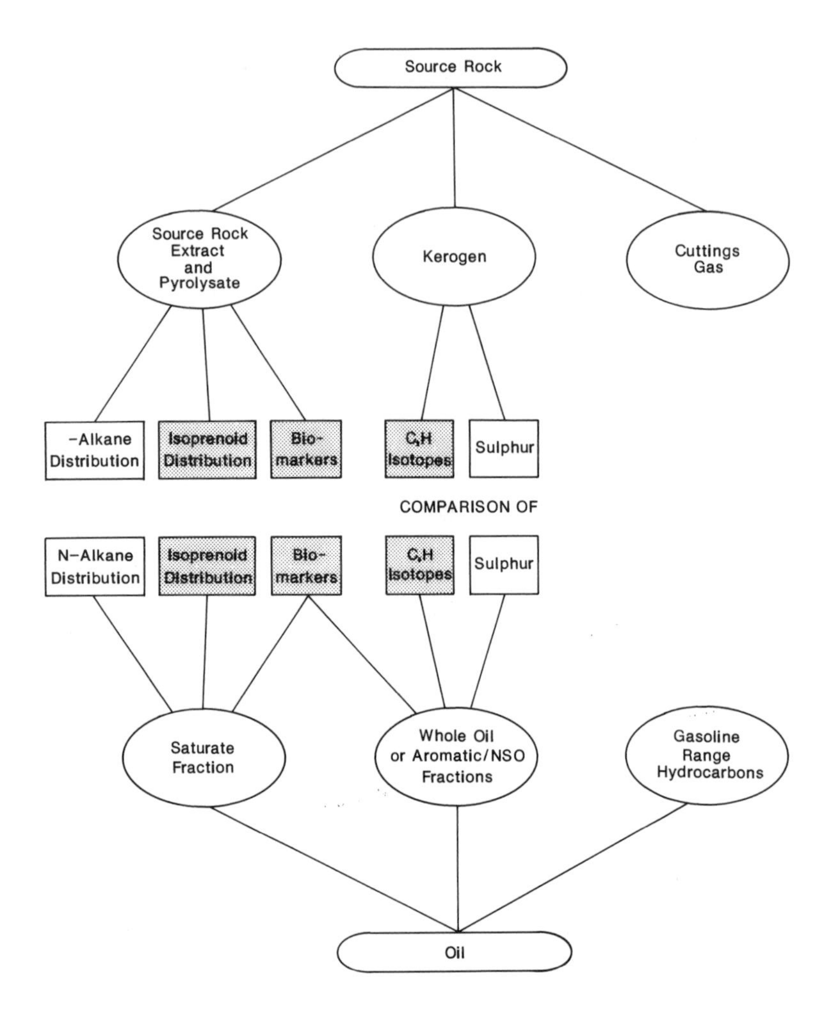

Gas characterization and gas–gas correlation

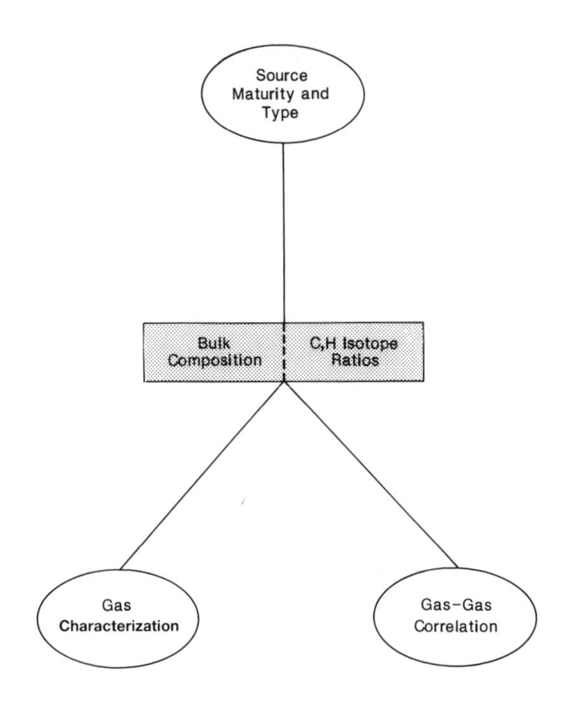

Reading list and general references

Basic chemistry texts

Chang, R. (1981). *Physical chemistry with applications to biological systems*, (2nd edn), 659pp. Macmillan Publishing Co., New York.

March, J. (1977). *Advanced organic chemistry: reactions, mechanisms and structure*, (2nd edn), 1328pp. McGraw Hill, New York.

Tedder, J.M., Nechvatal, A., Murray, A.W., and Cornduff, J. (1972). *Basic organic chemistry. Part 4: natural products.* J. Wiley, Chichester.

Basic petroleum geochemistry texts

Durand, B. (ed.), (1980). *Kerogen: insoluble organic matter from sedimentary rocks*, 519pp. Editions Technip, Paris.

Eglinton, G. and Murphy, M.T.J. (eds), (1969). *Organic geochemistry: methods and results*, 828pp. Longman, London.

Hunt, J.M. (1979). *Petroleum geochemistry and geology*, 617pp. W.H. Freeman, San Francisco.

Tissot, B.P. and Welte, D. (1984). *Petroleum formation and occurrence—a new approach to oil and gas exploration*, (2nd edn), 699pp. Springer-Verlag, Berlin.

Stach, E. (1982). *Stach's textbook of coal petrology*, (3rd edn), 535pp. (transl. D. Murchison, G. H. Taylor, and F. Zierke). Gebrüder Borntraeger, Berlin.

Applications of petroleum geochemistry

Brooks, J. and Welte, D.W. (eds), (1984). *Advances in petroleum geochemistry*, Vol. 1, 344pp. Academic Press, London.

Thomas, B. (ed.), (1985). *Petroleum geochemistry in exploration of the Norwegian Shelf*, 337pp. Norwegian Petroleum Society, Graham and Trotman, London.

Journals and periodicals

Bulletin of the American Association of Petroleum Geologists, monthly. Tulsa, Oklahoma.

Geochimica et Cosmochimica Acta, bi-monthly. Pergamon Journals, Oxford.

Nature, weekly. Macmillan Journals.

Organic Geochemistry, bi-monthly. Pergamon Journals, Oxford.

Glossary

α configuration The stereochemical abbreviation used primarily for *steroid and *terpenoid type molecules which is used to describe the configuration of a molecule in three dimensions. In two dimensions α is equivalent to into the plane of the paper. Shorthand form is ⁞ (see Fig.). It is conventional to list the carbon number at which the configuration occurs and the type of substituent group, e.g. *18α(*H*)-oleanane. It is paired with the *β configuration. The two forms represent *trans* and *cis* *geometrical isomers, αα and ββ being *cis* and αβ and βα being *trans*. An additional use of this notation is α, β, γ substitution on carbon chains, although this is not the most frequent use in *organic geochemistry. *See also*: isomer. *References*: Organic chemistry textbooks

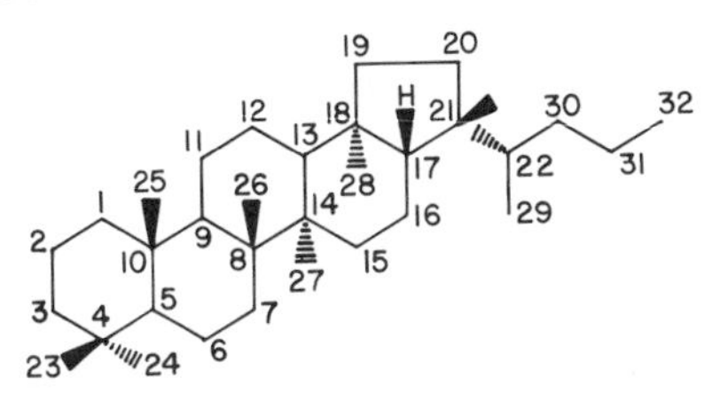

Hopane / Moretane

activation energy The energy which must be supplied to reacting substances to enable chemical transformation to take place is called the activation energy. It is measured in units of kilocalories per mol, or kilojoules per mol. The reaction goes through a transition stage during which an activated complex is formed (see Fig.). This has higher potential energy than either the reactants or products.

Apart from a basic relevance to all chemical transformations taking place in sediments, the activation energy has special significance in the field of petroleum geochemistry for estimating the degree of transformation of *kerogen into oil and gas, a technique known as *maturation modelling. Results of these models are applied in

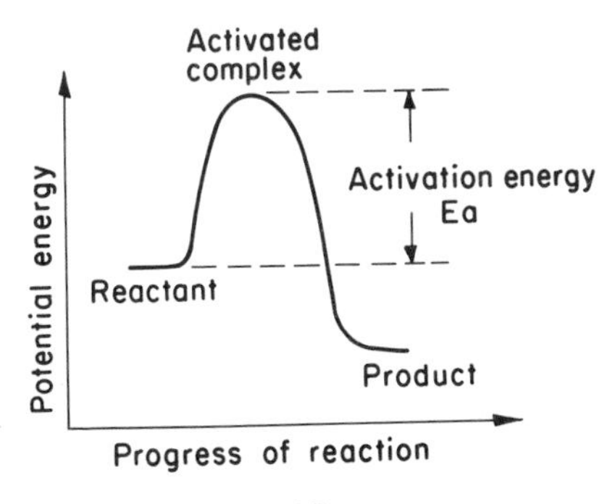

individual prospect and basin evaluations to assess the prospectivity before drilling, on the basis of the maturity of source rocks and the volumes of hydrocarbons produced. The activation energy, *A factor and temperature history are used in the *Arrhenius rate equation. The illustration given is for a single reaction. Kerogen is a complex and variable polymeric compound, which is capable of undergoing many different reactions simultaneously, hence a range of activation energies is probably more appropriate. This is supported by experimental observation; latest publications report values ranging from 45 to 80 kcals per mol.

The type and maturity of the kerogen determine the shape of the activation energy distribution. The results of these calculations, which are only predictions, are not real maturity measurements. If maturity measurements agree with model results, it is probable that the correct activation energies have been chosen for that site. Lateral variations in kerogen types in source rocks will, however, cause variations in activation energy distributions from site to site within a basin. Measurements related to activation energies may be obtained from *pyrolysis experiments. These have been informally called *pseudo-activation energies, as they are not true measurements of activation energies alone. The most realistic results are obtained if the pyrolysis is carried out in the presence of water. However, a method involving the use of *Rock Eval, which is anhydrous pyrolysis, has also gained acceptance. In order to obtain a representative range of activation energies, samples on which these measurements are determined must be immature. *See also*: A factor, Arrhenius rate equation, chemical kinetics, conversion index, maturation modelling, order of a chemical reaction, pseudoactivation energy, Tissot and Espitalié model. *References*: Physical chemistry textbooks.

acyclic isoprenoids Compounds which are composed of *isoprene units linked in branched or linear form are known as acyclic isoprenoids. Commonly encountered examples are *pristane, farnesane, and *phytane. They have a *diagnostic ion of 183. *See also*: isoprene, isoprenoids. *Reference*: Volkman, J.R. and Maxwell, J.R. (1984). In *Biological markers—a monograph*, (ed. R.B. Johns). Elsevier, Amsterdam.

adduction A separation technique based on molecular size. In *organic geochemistry this mainly involves the use of *urea and *thiourea, whose crystals link together to form hexagonal channels, similar to the structure of *zeolites. This enables molecules of a certain size to be trapped or adducted into the crystal lattice, thus causing separation of compounds in the same way as *molecular sieves. The most frequent use of adduction is in the removal of *normal alkanes from a *saturate fraction of an oil or extract, thereby increasing the relative concentration of cyclic *steranes and *triterpanes prior to *gas chromatography–mass spectrometry analysis. Synonymous with clathration and inclusion complexing. *See also*: molecular sieve, thiourea, urea.

aerobic Having free air or oxygen. Specifically applied in geochemistry to water column and near-surface sediment conditions with sufficient levels of dissolved oxygen to support *aerobic bacteria, approximately equivalent to >0.2 ml/l. Chemical and bacterial oxidation of organic and inorganic components can occur. In aerobic environments all but the most resistant organic matter is destroyed; oil and gas source rocks do not normally occur under these conditions. Aerobic conditions may prevail in any depositional environment, and tend to be the norm in open ocean and fluviatile settings. Synonymous with oxic. *See also*: anaerobic, dysaerobic.

aerobic bacteria Bacteria which use free oxygen directly from the atmosphere or from solution in water. They are found in surface and near-surface aeolian and sub-aqueous sediments. They are responsible for the majority of the destruction of organic matter produced in water and on land. *See also*: anaerobic bacteria.

aetio porphyrin *See* etio porphyrin.

A factor A constant of integration from the Arrhenius rate equation. Units are generally $seconds^{-1}$ or $years^{-1}$. A typical value would be in the order of 0.5×10^{18} s^{-1}. Synonymous with pre-exponential factor and frequency factor. *See also*: activation energy, Arrhenius rate equation, chemical kinetics, Tissot and Espitalié model. *References*: Physical chemistry textbooks.

albertite *See* solidified bitumen.

algae Primitive, dominantly aquatic plants, which were among the earliest life forms to evolve in the Precambrian. They may live in either *non-marine or marine aqueous environments, where they form a large proportion of phytoplankton. Algal remains are susceptible to bacterial and chemical degradation. Algae are rich in *lipids and when deposited in sediments under *anaerobic conditions are believed to be the precursors of many of the components of oils. They often contain large amounts of C_{17} *normal alkane, so its dominance in alkane distributions in oils can indicate an algal source. Marine brown and blue-green algae contain a dominance of C_{29} sterols and marine red algae C_{27} sterols. Non-marine blue-green algae contain dominantly C_{29} sterols. C_{28} sterols have become increasingly more abundant with geologic time in the marine environment. One of the *carotenoids, β-carotane has also been attributed to an algal source. As *kerogen, if they retain their shape they are called *alginite e.g. *Tasmanites* or **Botryococcus*, but they may be degraded sufficiently to become amorphous. In this form they are the precursor to the commonly encountered type of oil source material, *amorphous kerogen. They are hydrogen-rich and equate chemically to *Type I and *Type II kerogen. Latterly used synonymously with cyanobacteria. *See also*: alginite, amorphinite (1), amorphous kerogen, boghead coal, coorongite, torbanite.

alginite A term used in the microscopic study of *kerogen to describe the *liptinite *maceral which is composed of morphologically recognizable remains of unicellular or colonial *algae, commonly *Tasmanites* and *Botryococcus*. Its colour varies from yellow-green to black in transmitted white light, and it fluoresces green to yellow brown in UV light when immature to mature. It is a common constituent of *torbanites and *boghead coal. It is hydrogen-rich and equates chemically to *Type I kerogen (algal sapropel). *See also*: algae, *Botryococcus*. *Reference*: Allen, J., Bjorøy, M., and Douglas, A. (1980). A geochemical study of the exinite group maceral alginite selected from the Permo-Carboniferous torbanites. In *Advances in Geochemistry 1979* (ed. A. Douglas and J.R. Maxwell), pp. 599–618. Pergamon Press, Oxford.

aliphatic The term originally meaning fatty, but currently defined as organic compounds which are non-aromatic and non-polar, such as alkanes, alkenes, and alkynes, i.e. organic compounds which are open chain and those cyclic compounds which resemble open chain compounds in their chemical reactions. *References*: Organic chemistry textbooks.

alkanes *Carbon and *hydrogen compounds which are saturated, i.e. contain no carbon–carbon double or triple bonds. The simplest alkanes are *methane, CH_4, then *ethane, C_2H_6, and *propane, C_3H_8. The common suffix -ane indicates membership in the alkane series and a basic chemical similarity. Boiling points increase regularly, the first four or five alkanes being gases at normal temperatures, the next twelve are liquids. Alkanes from C_{17}+ are solids. They have low chemical activity but burn releasing large amounts of energy so they are important fuels. Alkanes fall into three structural groups, the *normal (or n-) alkanes which have straight carbon chains; the *branched alkanes, in which the carbon skeleton contains offshoots; and *cyclic alkanes (cyclo-), which have carbon atoms linked in a ring ranging from three members upwards. They are derived from practically all aqueous and *terrestrial organic matter, where they are part of the *lipid fraction. Alkanes can form from 5 to 80 per cent of crude oils and almost all of natural hydrocarbon gases. Crude oils which contain predominantly alkanes are economically attractive. Alkanes in oils form the major substrate for *bacteria and can be partially or totally removed if an oil suffers *biodegradation. An example of a normal alkane is normal pentane, C_5H_{12}, a branched alkane, pristane $C_{19}H_{40}$, and cyclic alkane, cholestane, $C_{27}H_{48}$ (see Fig.). Synonymous with saturate fraction,

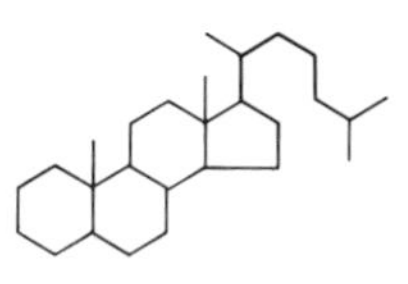

$C_{27}H_{48}$

Cholestane

paraffins, and naphthenes. *See also*: branched alkane, cyclic alkane, normal alkane, saturate fraction. *Reference*: Tissot, B., Pelet, R., Roucache, J. and Combaz, A. (1977). Alkanes as geochemical fossil indicators of geological environments. In *Advances in Organic Geochemistry 1975* (ed. R. Campos and G. Goni), pp.117–54. Enadimsa, Madrid.

alkenes *Carbon and *hydrogen compounds containing carbon–carbon double bonds are called alkenes. The common suffix -ene indicates membership of the series and basic chemical similarity. The simplest alkenes are ethene (ethylene, see Fig.), C_2H_4, and propene, C_3H_6. Alkenes are more reactive than alkanes. Structures are normal, branched, and cyclic. Biologically inherited alkenes undergo saturation during diagenesis to form a corresponding alkane. Alkenes are only minor components of *crude oils, *natural gases, and sediments. However they may be produced as decomposition products of oil-based drilling muds at high temperatures during *turbodrilling and during anhydrous *pyrolysis of *kerogen. Alkenes are refinery by-products of the cracking of oils. Synonymous with olefin. *References*: Organic chemistry textbooks.

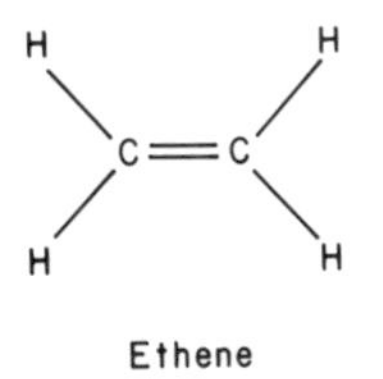

Ethene

alkynes Compounds with a carbon-carbon triple bond. These compounds are highly reactive and do not occur naturally in substantial amounts. The gas acetylene (ethyne, see Fig.) is an alkyne, C_2H_2. The suffix -yne indicates membership of the series.

$$H—C\equiv C—H$$

Ethyne

allochthonous organic matter Organic matter which is deposited some distance laterally from where it was formed before incorporation into a sediment is said to be allochthonous, e.g. *terrestrial plant material deposited in marine sediments. *Coal is called allochthonous or drifted when eroded from its place of formation and redeposited in other environments. The presence or absence of associated root beds beneath humic coals is an indicator of whether the coal is allochthonous or *autochthonous. The term may also be used to describe *vitrinite reworked from an older, more mature rock. *See also*: authochthonous organic matter, reworked kerogen.

amino acids Organic crystalline solids which contain *nitrogen and which are the building blocks of *proteins. They are amphoteric molecules containing a carboxyl group ($-CO_2H$) and an amine group ($-NH_2$). All amino acids except glycine contain a *chiral centre, and only the L forms are found in proteins. Most proteins comprise about 20 different amino acids linked in different ways. They are present in young sediments but their abundance decreases rapidly with depth. Amino acid isomerizations have been used to calculate thermal history data in young sediments. An example of an amino acid is isoleucine (see Fig.). *See also*: chemical kinetics. *Reference*: Hare, P.E. (1969). Geochemistry of proteins, peptides and amino acids. In *Organic geochemistry* (ed. G. Eglinton and M.T.J. Murphy). Springer-Verlag, Berlin.

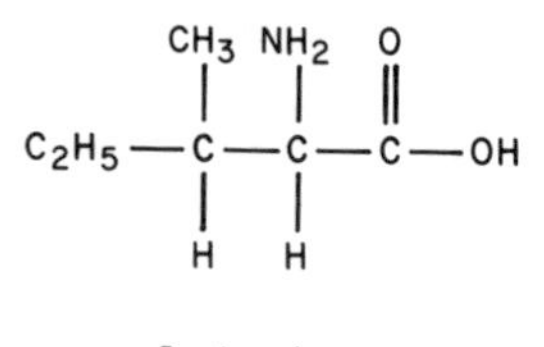

Isoleucine

amorphinite A non-standard, *coal petrographic term used to describe the maceral which is amorphous or lacking any distinct structure. Its colour, which is dependent on *maturation, is usually yellow to reddish brown in transmitted white light. This type of organic matter can be either oil prone or gas prone. It is identified by its appearance and by associated particles. Two types are defined.

(*a*) amorphinite l, is an oil prone material usually derived from phytoplankton and *algae. It is distinguished by its *fluorescence in UV light when *immature to *mature. Different types of *amorphous kerogen can be distinguished on the basis of fluorescence characteristics if not part of a mixture of different origins. It equates chemically with *Type I or *Type II kerogen

(*b*) amorphinite v, is gas prone and derived from bacterial degradation of humic material. It is reddish brown in transmitted white light and is distinguished by its dullness or lack of fluorescence in UV light. It may be confused with *post mature amorphinite l. It equates chemically with *Type III kerogen.

Differentiation between these two types is more reliable when the maturity levels are known and macerals have been examined in UV light. Before the routine use of UV light in organic petrography, many gas prone source rocks were incorrectly identified as oil prone. Synonymous with *sapropel (l), *amorphous kerogen, *unstructured liptinite (l), *liptodetrinite, and *vitrodetrinite. *See also*: amorphous kerogen, degraded vitrinite, unstructured liptinite.

amorphous kerogen *Kerogen which lacks distinct form or shape. The term often implies a type of oil prone material. However, wood and other humic-derived material, which is dominantly gas prone, can also become amorphous during bacterial degradation in surface and near-surface sediments. Amorphous material which fluoresces in UV light indicates a potential for oil, and is usually of algal or other phytoplanktonic origin. Lack of *fluorescence usually indicates a humic origin, unless the kerogen is at a high level of *maturity. Poor sample preparation may also reduce fluorescence. *Kerogen typing without the use of UV light cannot reliably distinguish between oil and gas prone material, so interpretation of hydrocarbon potential should be made cautiously and supported by chemical analysis, e.g. *Rock Eval pyrolysis. Synonymous with *sapropel, *amorphinite (l or v), *unstructured liptinite, degraded vitrinite, *liptodetrinite, and *vitrodetrinite; it may equate chemically with *Type I, *Type II, or *Type III kerogen.

anaerobic Without free oxygen. The term is dominantly applied in organic geochemistry to water column, surface and near-surface sedimentary environments, with insufficient oxygen to support *aerobic bacteria. Less resistant organic matter, which would normally be destroyed by chemical oxidation, aerobic and *anaerobic bacteria, may be preserved in anaerobic environments.

Anaerobic conditions can occur in any sedimentary environment, but most frequently in very deep water, swamps, lakes or ponds, silled or restricted basins, or in shelf settings. They may be induced in a previously *aerobic setting, when aerobic bacteria use up available oxygen, which is not replenished. High sedimentation rates may cause anaerobic conditions by cutting off sediments quickly from oxygenated bottom water. Anoxia may be produced by the establishment of haloclines or thermoclines, especially in *lacustrine environments. Totally anaerobic conditions are rare, as there is always some aeration of near-surface water. Sediments deposited under anaerobic conditions tend to be laminated as there are no benthonic fauna to destroy natural laminations, as happens in more ventilated settings. Anaerobic environments are prime sites for the deposition of source rocks for both oil and gas. The *pristane/phytane ratio of oils and source rock extracts is often used to estimate whether the source depositional environment was anaerobic or aerobic. The pristane/phytane ratio is lower in anaerobic environments.

There are plate tectonic controls which make anaerobic environments more frequent at certain times in the sedimentary record, the Jurassic and Cretaceous being examples of such times. The breakup of Pangaea caused many narrow restricted basins. In the North Atlantic, this resulted in marine transgression and widespread Liassic source rocks. Later, widespread restricted marine conditions were again established in the Late Jurassic with the deposition of more source rocks, e.g. the Kimmeridge Clay and Draupne Formations of the north-west European oil provinces. The same series of events occurred in the South Atlantic but later, in the Cretaceous. Marine transgression failed to reach some South American/African coastal basins so that lakes formed and early source rocks are *non-marine. Later, the well-known marine Cretaceous black shale sequences were deposited.

The term anaerobic is applied to environments where there is <0.1 ml/l of dissolved oxygen. Synonymous with euxinic, reducing, and anoxic. *See also*: aerobic, dysaerobic. *Reference*: Demaison, G. and Moore, G. T. (1980). Anoxic environments and oil source bed genesis. *Am. Assoc. Pet. Geol. Bull.*, **64**, 1179–209.

anaerobic bacteria Anaerobic bacteria are those which obtain their *oxygen from the reduction of oxygen-containing compounds rather than from free oxygen dissolved in water. In *organic geochemistry, *sulphate-reducing bacteria are the most important in the *diagenesis of organic matter (see Fig.). Framboidal *pyrite in sediments is indicative of the action of anaerobic bacteria. *See also*: methanogenic bacteria, sulphate-reducing bacteria.

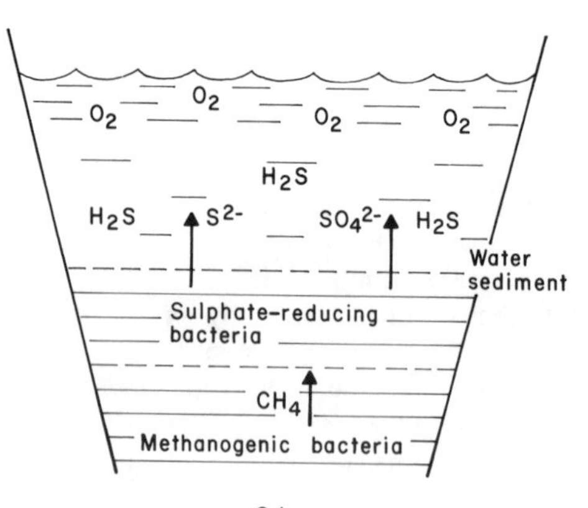

anhydrous pyrolysis *See* pyrolysis.

anoxic *See* anaerobic.

anteiso-alkanes Straight chain *alkanes with a methyl carbon branch on the third carbon; shorthand 3-methyl alkanes, e.g. 3-methylhexane or methylhex-3-ane (see Fig.). *References*: Organic chemistry textbooks.

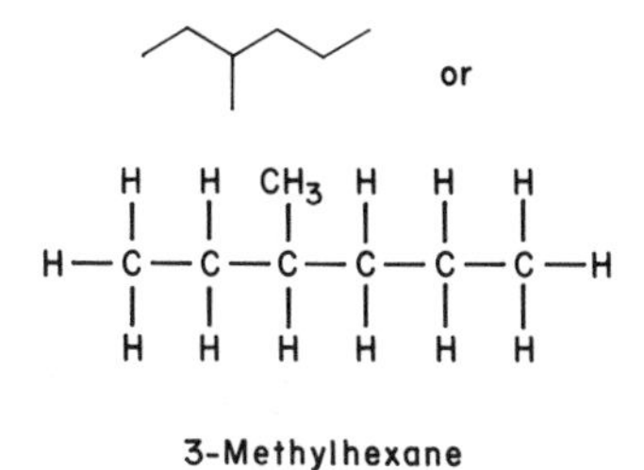

3-Methylhexane

antimer *See* enantiomer.

API American Petroleum Institute. The API has many functions. It has standardized various terms, analytical techniques, and measurements in the petroleum industry, e.g. 'API method for *asphaltene analysis'.

API gravity American Petroleum Institute measure of the specific gravity (s.g.) of *crude oils and *condensates. The specific gravity is converted to an integer by use of the formula:

$$\frac{141.5}{\text{s.g. at }60°\text{F or }16°\text{C}} - 131.5.$$

This integer is easier to display and to use than a four-place decimal number. The API gravity increases with decreasing specific gravity. Examples are: s.g. 0.6501, API gravity 86°; 0.7603, 55°; 0.8705, 31°; 0.9505, 17°. A heavy crude oil is defined as one which has an API gravity of $<25°$, a medium gravity crude 25° to 35°, a light crude 35° to 45° and a condensate $>45°$. Fresh water has an API gravity of 10°. API gravity may also be estimated from oil fluorescence colour in UV light. Brown fluorescence indicates an API of $<15°$; orange, 15° to 25°; cream/yellow 25° to 35°; white 35° to 45°; and blue/white to violet $>45°$. *Biodegradation and *water-washing decrease the API gravity of oils.

arboranes *See* triterpanes.

aromatic compounds Originally meaning 'fragrant' compounds, but now defined as *benzene and those compounds which resemble benzene in their chemical behaviour. They are unsaturated, and are represented as containing one or more rings with conjugated (alternating) carbon–carbon double and single bonds; in reality these are delocalized and form an orbital ring current. Simple aromatic compounds are responsible for *fluorescence in oils and *extracts. An example of an aromatic compound is *naphthalene (see Fig.).

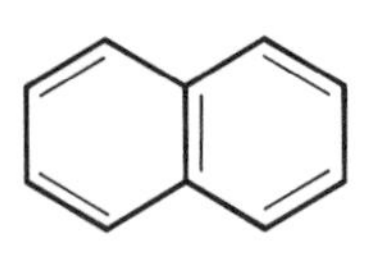

Naphthalene

aromatic fraction That fraction of *crude oils and rock *extracts which is usually eluted with *benzene, *toluene, or mixtures containing these solvents, during *liquid chromatography or *HPLC, expressed as a weight per cent. It contains the low to medium molecular weight aromatic *hydrocarbons. The quantity and type of aromatic compounds in crude oils or extracts is determined both by maturity and the type of source organic matter; aromatic compounds being more prevalent in *humic organic matter. The aromatic fraction decreases in oils and extracts on increasing maturity due to cracking, whilst residual *kerogen becomes more aromatic. The greater solubility of low molecular weight aromatic compounds in water means that they may be preferentially removed during *water-washing of oils. The aromatic fraction of oils is generally little affected by *biodegradation. *See also*: saturate to aromatic ratio.

aromatic steroids Aromatic compounds found in oils and sediment *extracts which are probably derived from sterols via sterenes with two double bonds. Different families of aromatic steroids result from the aromatization of different rings, e.g. C ring or A ring, and they may be rearranged or regular, as with steranes. Aromatic steroids may be identified by *gas chromatography–mass spectrometry (GC–MS) using *m/e* 143 and 231 for triaromatics, 239 for demethylated monoaromatics, 245 for methyl triaromatics, and 253 for monoaromatics (see Fig.).

Aromatic steroids are used for *oil to oil and *oil–source rock

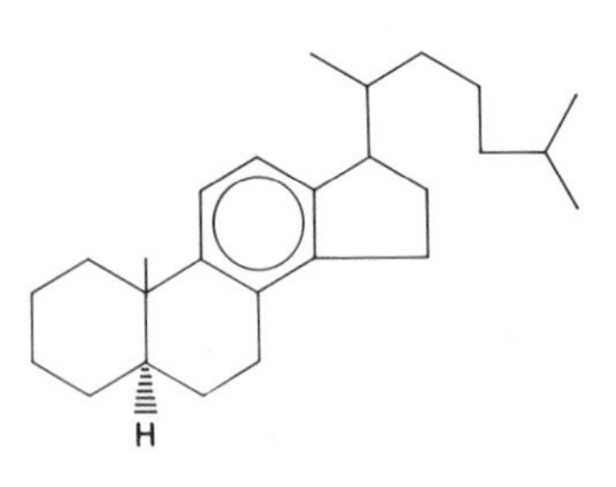

C_{27} Monoaromatic steroid

correlation, especially in oils which have undergone *biodegradation. Source-sensitive parameters are: rearranged to regular steroid distributions, 5β- to 5α-methyl steroid distributions, the relative abundance of C_{26} to C_{29} mono- and triaromatic isomers, and relative abundance of methyl triaromatic isomers. The abundance of 5α-methyl rearranged steroids may be indicative of sabkha environments. The relative abundance of C_{27}:C_{28}:C_{29} monoaromatic steroids may be a more reliable marine:non-marine indicator than for regular steranes.

Progressive aromatization occurs as a result of *maturation so that the ratio of one ring (mono)aromatic to three ring (tri)aromatic steroids is a maturation parameter. This reaction has also been used in *inverse modelling of basin *heatflow, with isomerization reactions. Another maturation parameter is derived from the progressive cracking of side chains of the aromatic steroids. *See also*: chemical kinetics, inverse modelling. *Reference*: Riolo, J., Hussler, G., Albrecht, P., and Connan, J. (1986). Distribution of aromatic steroids in geological samples: their evaluation as geochemical parameters. *Org. Geochem.*, **10**, 981–90.

Arrhenius rate equation Chemical reactions occur at different rates at different temperatures. The *rate constant for a reaction varies with temperature. This variation is described by the Arrhenius rate equation. It is derived from the Van t'Hoff equation. In a balanced reaction where the rate constant of the forward reaction is k_1 and the reverse reaction is k_2, the variation of the equilibrium constant $K(k_1/k_2)$ is given by:

$$\frac{\mathrm{d}(\ln K)}{\mathrm{d}T} = \frac{Q}{RT^2},$$

where Q is the heat of reaction, R is the gas constant, T the temperature in degrees Kelvin.

Since $\ln K = \ln k_1 - \ln k_2$

$$\frac{\mathrm{d}(\ln k_1)}{\mathrm{d}T} - \frac{\mathrm{d}\ln k_2}{\mathrm{d}T} = \frac{Q}{RT^2}$$

Therefore,

$$\frac{\mathrm{d}(\ln k_1)}{\mathrm{d}T} = \frac{E_{a1}}{RT^2} + \mathrm{A}$$

and

$$\frac{\mathrm{d}(\ln k_2)}{\mathrm{d}T} = \frac{E_{a2}}{RT^2} + A,$$

where $Q = E_{a1} - E_{a2}$ and E_a is the activation energy.

Arrhenius found that the variation of k with temperature could be expressed satisfactorily by the simplified equation:

$$\frac{d(\ln k)}{dT} = \frac{E_a}{RT^2}$$

or

$$\ln k = \frac{-E_a}{RT} + \ln A.$$

The integrated form of the Arrhenius rate equation is:

$$k = A \exp(-E_a/RT),$$

where k = rate constant, and A = a constant called the frequency factor or *A factor. This equation, although not totally accurate, is adequate for the majority of chemical reactions.

One of the applications of the Arrhenius rate equation to geochemistry is in the modelling of *kerogen conversion to oil and gas, if the temperature history, activation energy, and A factor of the source rock (and hence the kerogen) are known. The conversion of kerogen is assumed to be a *first-order reaction. Use of this equation in various accurate or approximate forms is the basis of *theoretical maturity calculations for *maturation modelling. *See also*: activation energy, A factor, chemical kinetics, Lopatin, maturation modelling, Tissot and Espitalié model. *References*: Physical chemistry textbooks.

asphalt The *API defines asphalt as a dark brown, viscous liquid or low-melting solid that consists dominantly of *hydrocarbons, and is soluble in carbon disulphide, or insoluble in normal heptane. Asphalt may be produced commercially from the distillation of crude oil or naturally by the weathering of shallow hydrocarbon accumulations, e.g. Trinidad. Synonymous with pitch, tar, *asphaltenes, and *bitumen. *See also*: solidified bitumen. *Reference*: Chilingarian, G.V. and Yen, T.F. (1978). Bitumens, asphalts and tar sands. In *Developments in Petroleum Science*, vol. 7. Elsevier, Amsterdam.

asphaltenes The heavy molecular weight component of crude oils or sediment *extracts which is soluble in carbon disulphide or insoluble in normal heptane. Chemically they are polyaromatic nuclei linked by *aliphatic chains or rings and functional groups (see Fig.). The asphaltene fraction of oils is traditionally determined prior to *liquid chromatography. However, *saturate fraction compounds may be occluded with the asphaltenes. To avoid this, they can be precipitated from the *NSO/polar fraction after the saturate fraction has been removed. The content of asphaltenes is conventionally expressed as a percentage by weight of oil or extract in a rock. The molecular weight of individual asphaltene molecules can be from 500 to 1500. Groups of molecules called micelles form, which can increase the molecular

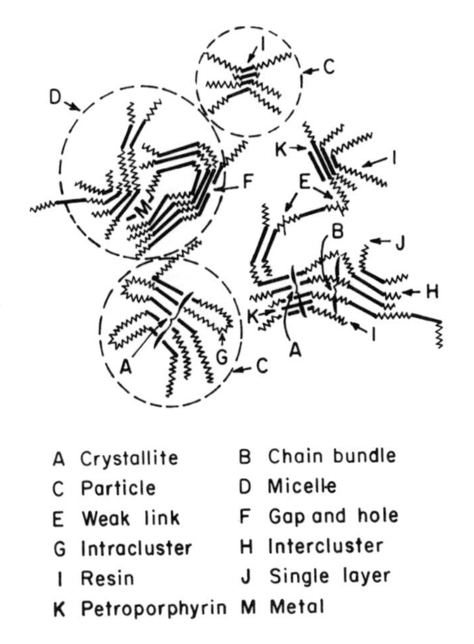

weight to several thousand. The illustration shows a schematic representation of how various parts of an asphaltene molecule may be linked together.

Asphaltenes may be produced by the action of heat, *bacteria, or UV light and are in solution in oil. They are precipitated by gas solution and may form a *tar mat at the base of an oil column. On average they contain 12 per cent by weight of nitrogen, sulphur, and oxygen compounds. *Elemental analysis or *stable carbon isotope ratios may enable the various modes of formation of asphaltenes to be discriminated. Severe thermal maturation will give asphaltenes with an *H/C ratio of <0.53; higher values are obtained from asphaltenes precipitated by gas solution. *Pyrolysis of asphaltenes gives a product of similar character to the saturate fraction of parent oils. Use of this may be made when correlating heavily biodegraded oils with source rocks or other oils. *Reference*: Pelet, R., Behar, F., and Monin, J.C. (1986). Resins and asphaltenes in the generation and migration of petroleum. *Org. Geochem.*, **10**, 481–98.

autochthonous organic matter The term autochthonous is usually used to denote organic matter originating in, or over, the water column above a depositional site, implying little or no horizontal transport. Rooted or well-bedded *coals would be obvious examples of autochthonous material. The term is also used by some analysts to denote the *vitrinite reflectance population which is considered to be truly representative of the maturity of a sample, synonymous with *in situ* vitrinite. *See also*: allochthonous organic matter.

β configuration The stereochemical abbreviation used primarily for *steroid and *terpenoid type molecules which is used to describe the configuration of a molecule in three dimensions. In two dimensions β is equivalent to out of the plane of the paper or above the plane of the molecule. Shorthand form is ▼ (see Fig.). Paired with the α configuration, the two forms represent *trans* and *cis* *geometrical isomers, αα and ββ being *cis* and αβ and βα being *trans*. An additional use of this notation is α, β, γ substitution of carbon chains, although this is not the most frequent use in *organic geochemistry. *See also*: isomer. *References*: Organic chemistry textbooks.

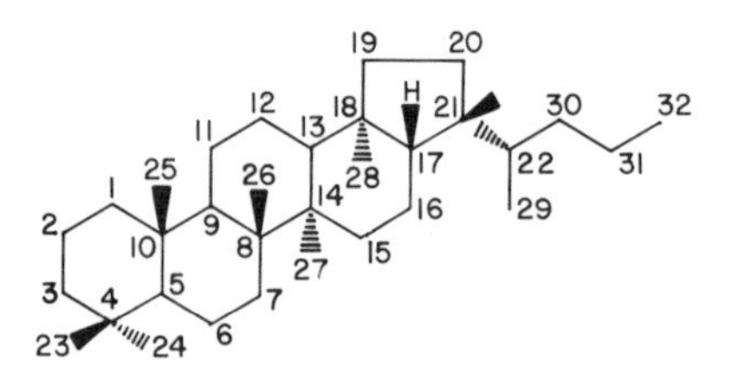

Hopane / Moretane

bacteria Uni- or multicellular, prokaryotic life forms which have been preserved in the sedimentary record since the Precambrian. They are subdivided into *aerobic bacteria which metabolize oxygen from the atmosphere or from solution in water, and *anaerobic bacteria, which derive oxygen from the breakdown of oxygen-containing compounds, e.g. *sulphate-reducing bacteria, or *methanogenic bacteria.

Bacteria play an important role in the degradation of plant material during early *diagenesis, both above and below the sediment–water interface. Bacteria may also play a large part in the alteration of oil characteristics if introduced by ground water after accumulation, or when oil migrates to shallow locations, when *biodegradation may occur. Bacteria are a source of the *hopane series of *triterpanes, *methyl steranes, C_{27} to C_{29} *steroids, and other C_{10} to C_{30} hydrocarbons found in oils and sediment *extracts. *See also*: biodegradation, methanogenic bacteria, sulphate-reducing bacteria.

balkashite *See* coorongite.

base peak Dominant fragment produced during *gas chromatography–mass spectrometry (GC–MS). *See also*: diagnostic ion.

benzene The simplest *aromatic compound, containing one, six-membered carbon ring with conjugated double and single carbon–carbon bonds, formula C_6H_6 (see Fig.). It is present in all *crude oils except those which have been weathered, with resulting loss of light hydrocarbons and *gasoline range hydrocarbons, or those in which *water-washing has taken place, as it is one of the more water-soluble hydrocarbons.

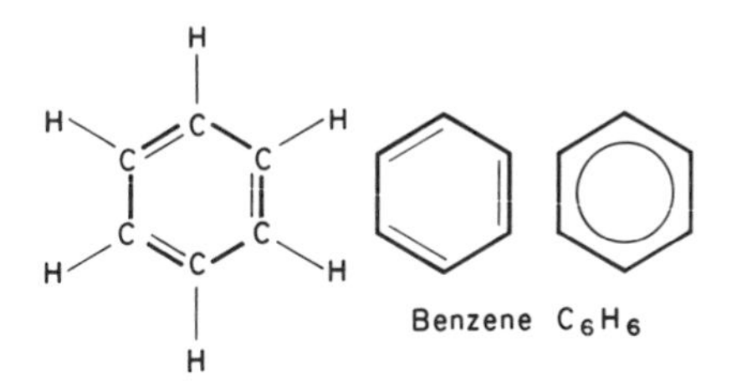

Benzene C_6H_6

biodegradation This term is normally applied to the degradation of oil by *bacteria, which may occur when an accumulation is unroofed, when seeps of oil reach the surface or near-surface, or when ground water is introduced into an accumulation by hydrodynamic flow. *Aerobic or *anaerobic bacteria are introduced into the oil, parts of which are a nutrient for bacteria. Although different types of bacteria may attack different portions of the oil, generally the *normal alkanes are the first compounds to be removed (see Fig., p. 32), *branched alkanes follow, then eventually *cyclic alkanes (see Table p. 33). *Demethylated hopanes and demethylated *tricyclic terpanes have been identified from many biodegraded oils. These are probably the result of enzyme attack on *hopanes. Only rarely are *aromatic compounds attacked, as they are generally poisonous to bacteria.

Oils which have been biodegraded may show fairly distinct compositional and *stable carbon isotope trends. The *asphaltene fraction increases as a result of bacterial attack. (These asphaltenes have a lighter isotopic ratio than asphaltenes from undegraded oils from the same source rock.) *API gravity normally decreases; viscosity increases, except in waxy oils where the viscosity may decrease due to the removal of long-chain alkanes. Bacterial activity will cease when the temperature in the reservoir increases. A value of 60°C or 140 to 190°F is normally considered sufficient to, effectively, end activity. *See also*: water-washing. *Reference*: Alexander, R., Kagi, R.I., Woodhouse, G.W., and Volkman, J.K. (1983). The geochemistry of some biodegraded Australian oils. *APEA J.*, **23**, 53–63

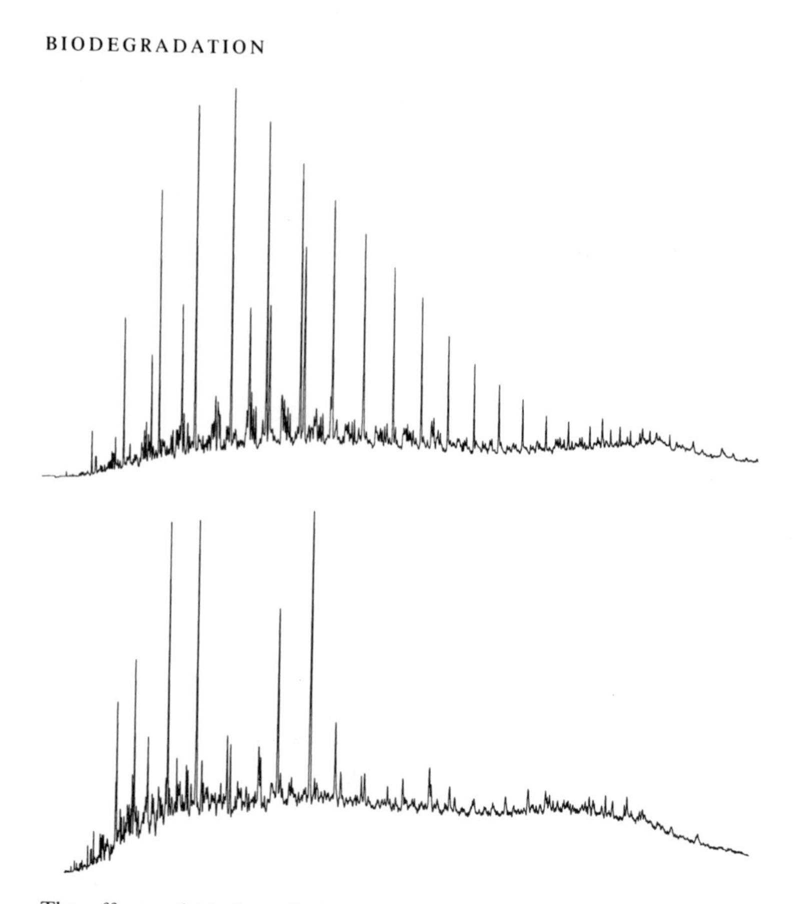

The effects of biodegradation on the composition of a typical mature paraffinic oil (from Alexander *et al.* 1983).

Level	Chemical composition	Extent of biodegradation
1	Typical paraffinic oil; abundant n-alkanes	Not degraded
2	Light end n-alkanes removed	Minor
3	>90% n-alkanes removed	
4	Alkylcyclohexanes removed; isoprenoids reduced	Moderate
5	Isoprenoid alkanes removed	
6	C_{14}-C_{16} Bicyclic alkanes removed	Extensive
7	>50% (20*R*)-5α, 14α, 17α(*H*)-steranes removed	Very extensive
8	Steranes altered; demethylated hopanes abundant	Severe
9	Demethylated hopanes predominate; some change to diasteranes; no steranes	Extreme
10	20*S*-13β(*H*), 17α(*H*) diasteranes degraded more rapidly than 20*R* epimers	

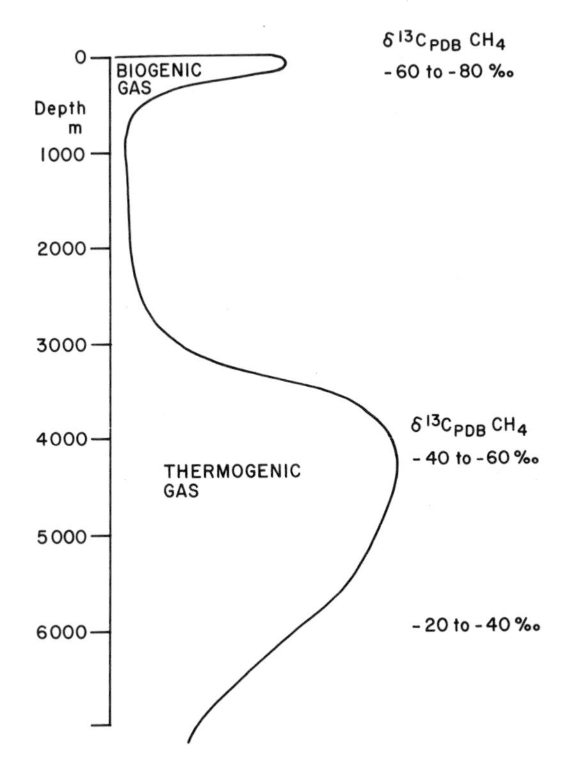

biogenic gas Methane which is produced by the action of *bacteria on organic matter at shallow depths (several metres) is called biogenic gas, and is commonly known as swamp or marsh gas. Compositionally it is a very *dry gas, being normally 99.9 per cent methane. It also has a distinctive stable carbon isotopic composition of −50 to −80 per mil *PDB (20–40 per mil depleted in ^{13}C). It normally escapes to surface and is lost to the atmosphere (see Fig.), but in areas with high sedimentation rates it may be preserved as shallow gas accumulations. Complexed with water under a narrow range of temperatures and pressures, methane may form *gas hydrates, which may be identified on seismic sections as bottom-simulating reflectors. Shallow gas accumulations may be a drilling hazard, but can sometimes be detected as bright spots on seismic sections. Large accumulations may be commercially exploitable, e.g. Po Valley, Italy; Rice, USA; Siberia, USSR. *See also*: methane, methanogenic bacteria, dry gas, gas hydrate. *Reference*: Rice, D.D. and Claypool, G.E. (1981). Generation, accumulation and resource potential of biogenic gas. *Am. Assoc. Petr. Geol. Bull.*, **65**, 5–25.

biological marker (biomarker) Compounds, or characteristics of compounds, found in petroleum or rock *extracts which indicate an unambiguous link with a natural product are known as biological markers. Diagenetic changes which occur in a sediment may alter functional groups and bonds in the natural product but the carbon skeleton of the compound remains the same. The simplest compounds which are biological markers are *normal alkanes derived from land plant *waxes and *fatty acids, *isoalkanes, and *isoprenoids. *Chlorophyll decomposes to *porphyrins and to *pristane and *phytane from the side chain.

Optically active compounds also indicate a biological origin. *Steranes and *triterpanes in oils and *extracts are common biological markers and are derived from *steroids and *terpenoids. Synonymous with geochemical fossils and molecular fossils. *See also*: isoprenoids, optical isomers, porphyrin, sterane, triterpane. *Reference*: Seifert, W.K. and Moldowan, J.M. (1981). Palaeoenvironmental reconstruction by biological markers. *Geochim. et Cosmochim. Acta*, **45**, 783–94.

bisnorhopane The C_{28} *hopane with two methyl groups missing relative to C_{30} hopane (see Fig.). Generally these two groups are absent from the C-28 and C-30 positions, although a C-29 and C-30 'nor' compound has also been identified. It is relatively uncommon in oils and *extracts and is hence a good *oil-source correlation parameter. Its source is unknown, but was originally thought to be from ferns. There is also comment in the literature that it may be derived directly from *anaerobic bacteria but there is no chemical evidence to support this.

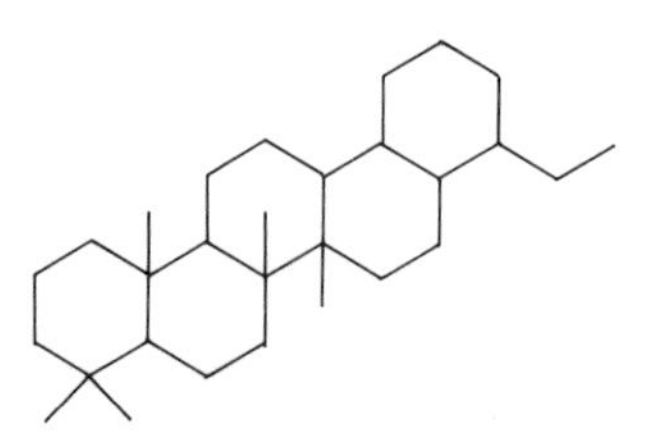

$17\alpha(H)$, $18\alpha(H)$, $21\beta(H)$–28, 30–bisnorhopane

bitumen A well-used but ill-defined geochemical term which has several meanings. It can be that fraction of *organic matter in sedimentary rocks which is soluble in *organic solvents (*extract). It may be the precursor to oil, or be the substance which occurs naturally in veins, pools, or dispersed in reservoir or source rocks. It is also used synonymously with organic matter which may be thermally extracted from rocks, and used informally to mean tar, pitch, and asphalt. *See also*: exsudatinite, extract, solidified bitumen.

bitumen/total organic carbon *See* extract to organic carbon ratio.

bituminite The *maceral name for *liptinite of uncertain but probable algal origin, found in brown *coals and *source rocks. It lacks a definite shape, has weak *fluorescence in UV light, which increases with length of exposure. It is sometimes used synonymously with *exsudatinite but is correctly a primary form of *kerogen.

boghead coal Sapropelic coals which are composed dominantly of algal remains are known as boghead coals. They differ from *humic coals in appearance, as they are black, dull, and have a conchoidal fracture. Deposits are usually thin and not laterally extensive. They produce oil when thermally mature but, owing to their limited volumes, do not usually constitute commercial oil sources. They are common in the Carboniferous of the Midland Valley of Scotland (*torbanites) and the Permian of France. They equate chemically with *Type I kerogen. *See also*: sapropel.

boghead peat *See* coorongite.

botryococcanes A group of long-chain *irregular *isoprenoid *hydrocarbons formed from botryococcene, found in **Botryococcus braunii*. Originally only one compound was discovered, formula $C_{34}H_{70}$ (see Fig.), but now several of these types of compounds have been identified. When found in oils they are a specific environmental indicator. They may be identified by *gas chromatography or *gas chromatography–mass spectrometry, with a *diagnostic ion of 113, and *molecular ion of original compound 478. *Reference*: Moldowan, J.M. and Seifert, W.K. (1980). The first discovery of botryococcane in petroleum. *J. Chem. Soc. Chem. Commun.*, **912**, 714.

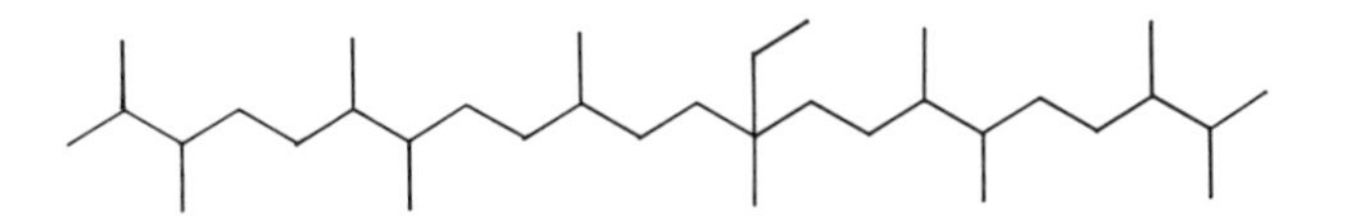

Botryococcane

Botryococcus (braunii) A fresh or brackish water, colonial alga which has apparently been present since the Precambrian. There are two races, A and B, with B giving rise to *botryococcanes and A to non-isoprenoid hydrocarbons. Thick, oily, gelatinous, decay-resistant deposits of *Botryococcus*, commonly 5 to 10m thick, may accumulate on lake bottoms and compact into *boghead coals. It is an excellent source for waxy oils although deposits are usually thin and only locally developed. Equates chemically with *Type I kerogen. *See also*: coorongite. *Reference*: Largeau, C., Casadevall, E., Kadouri, A., and Metzger, P. (1984). Formation of Botryococcus-derived kerogens—comparative study of immature torbanite and of the extant alga *Botryococcus braunii*. *Org. Geochem.*, **6**, 327–32.

branched alkanes Alkanes which have a branched carbon skeleton, including *isoalkanes, *anteiso-alkanes, and *isoprenoid alkanes. They are derived from many sources.

burial history The burial of one or more horizons traced through time, from deposition to (commonly) present day, is known as a burial history (see Fig.). The data are usually represented graphically as depth-time plots. These are used in conjunction with thermal models to estimate the temperature and hence *maturation history of source rocks in *maturation modelling. They do not usually take into account water depth at time of deposition and sedimentary loading effects on basement subsidence. *See also*: subsidence history.

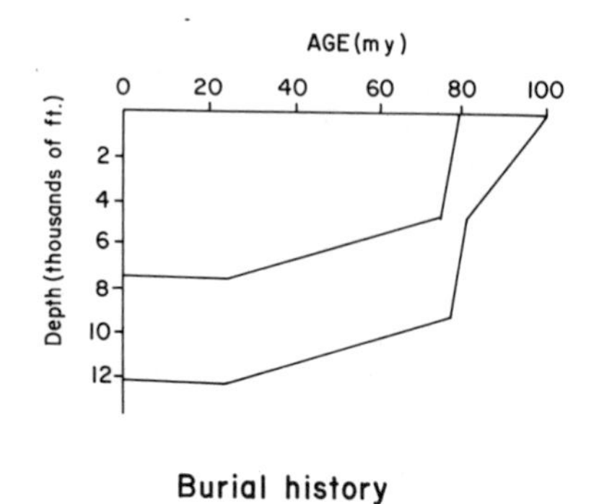

Burial history

butane The *normal and *isoalkane containing four carbon atoms, formula C_4H_{10} (see Fig.). Butane is a gas at normal temperatures and pressures. Both normal and isobutane occur in most *hydrocarbon accumulations, with the exception of *biogenic gas. The gas is analysed in mud, *headspace gas analysis and *cuttings gas analysis. The ratio of iso- to normal butane has been used as a *maturation indicator. The ratio is high in *immature sediments but decreases to a value of ~0.4 to 1.0 (commonly taken as equivalent to the onset of oil generation), but it is not a reliable parameter. *Biodegradation may change the $i\text{-}C_4/n\text{-}C_4$ ratio. *Reference*: Alexander, R., Kagi, R.I., and Woodhouse, G.W. (1983). Variation in the ratio of isomeric butanes with sediment temperature in Western Australia. In *Advances in Organic Geochemistry 1981* (ed. M. Bjorøy *et al.*), pp. 76–9. John Wiley, Chichester.

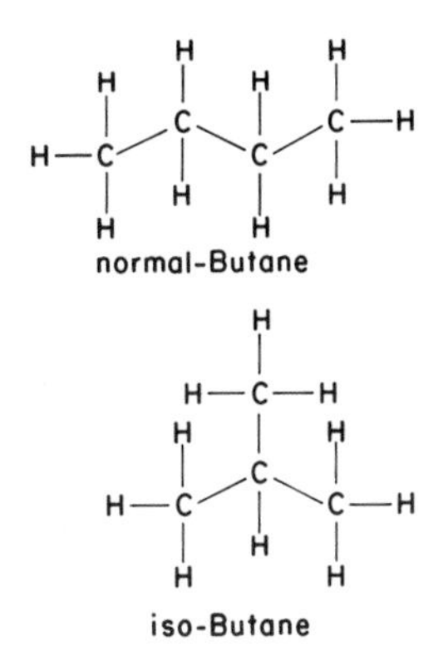

C_{15}+ The abbreviation used to describe the *saturate fraction of oils and *extracts, so called because after topping of oils at 200°C or extraction of source rocks, volatile hydrocarbons below C_{15} are mostly absent.

cannel coal Black, dull, sapropelic coal with a greasy lustre and conchoidal fracture. It is very similar to *boghead coal in appearance but strictly contains no *alginite; however there are transitional coals with both. They are rich in *sporinite and occur as seams within *humic coals.

Canon Diablo meteorite triolite *See* sulphur isotopes.

carbohydrates Research on compounds manufactured by cells indicated that they contained C, H, and O in the ratio of 1:2:1. This ratio was interpreted as a molecule of water bonded to an atom of carbon—hydrated carbon. The compounds were hence called carbohydrates. Subsequent research has shown that this is not true but the name has remained. The current understanding is that carbohydrates are polyhydroxy aldehydes and ketones, or substances which yield these compounds when hydrolysed. There are three major categories: monosaccharides or simple sugars, oligosaccharides with two to ten monosaccharides, and polysaccharides which are polymers of up to 500 monosaccharide units. Carbohydrates decompose to water-soluble sugars in near-surface sediments. They are constituents of almost all types of *organic matter. Cellulose is a carbohydrate (see Fig., p. 38). *References*: Organic chemistry textbooks.

carbon The element with symbol C and atomic weight 12. Carbon is a solid at normal temperatures and has two crystal forms, diamond and

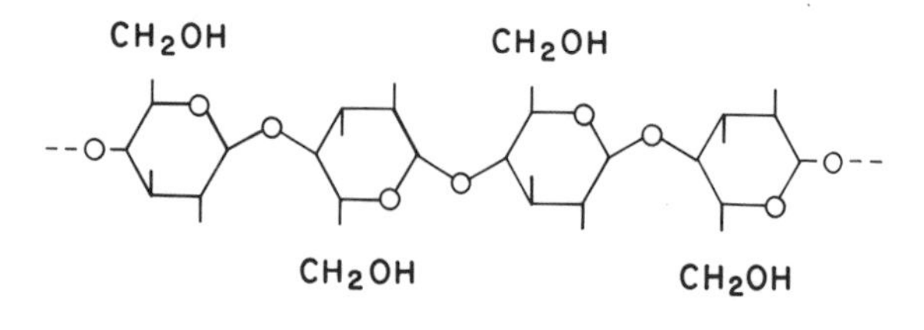

Cellulose

graphite. *Organic matter degrades to graphite by loss of H, O, N, and S at high levels of maturity. Carbon has three *isotopes, stable ^{12}C and ^{13}C, and unstable ^{14}C. It forms the skeleton of organic compounds, and its ability to form chains and rings enables it to form an immense variety of compounds. The study of the chemistry of its compounds is called organic chemistry.

carbon cycle The turnover of organic carbon in plants, animals, and rocks as a continuous cycle of respiration, photosynthesis, growth, decay, and preservation in sediments is called the carbon cycle (see Fig.). The cycle has two parts, which have different frequencies. The short part is the surface cycle of carbon through the atmosphere, hydrosphere, and biosphere, and back to the atmosphere. This part takes from days up to thousands of years. The second part is the cycle of carbon which is trapped in rocks, which is returned to the first cycle after a period of millions of years via the generation of oil and gas and weathering of carbon residues from rocks. Both parts of the cycle have occurred throughout geological time, although the amount and type of *organic matter has changed since the Precambrian.

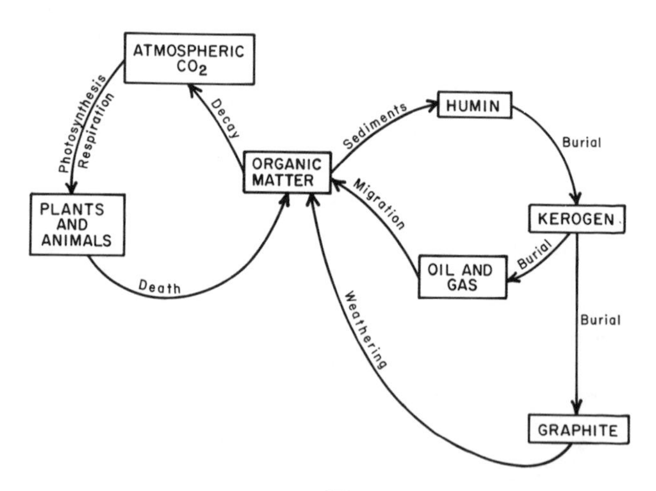

carbon dioxide Compound, formula CO_2, which is a gas at room temperature. It is the fundamental form in which carbon is cycled through plants, animals, and carbonates.

carbon isotopes *Carbon exists in three isotopic forms, the abundant ^{12}C, which is stable, ^{13}C which is stable but present in much smaller amounts than ^{12}C, and unstable or radioactive ^{14}C. The half life of ^{14}C is 5730 years; it is used for radiocarbon dating of organic artifacts and archaeological material of up to 30 000 years old.

In organic matter, the abundance of ^{13}C relative to ^{12}C depends on the metabolic pathway of carbon dioxide in the plant or animal, e.g. green plants use *photosynthesis and animals respire. This ratio is measured relative to a standard, the most common being the *PeeDee Belemnite, or *PDB. This isotopic contrast is not destroyed, but may be modified when sediments become lithified or when plant material breaks down and eventually forms kerogen. Consequently, environmental information may be obtained from organic matter in sedimentary rocks. The pre-Carboniferous isotopic composition was, on average, 4 per mil lighter than in post-Carboniferous organic matter. *See also*: stable carbon isotopes.

Carbon Preference Index The ratio of the abundance of odd carbon number *normal alkanes to even number normal alkanes measured from *gas chromatography of the *saturate fraction of an oil or *extract (see Fig.) is known as the Carbon Preference Index (CPI). Normal alkanes derived from *fatty acids, alcohols, esters, and land plant leaf *waxes, frequently give a predominance of odd-numbered alkanes in shallow sediments. This predominance decreases with increasing

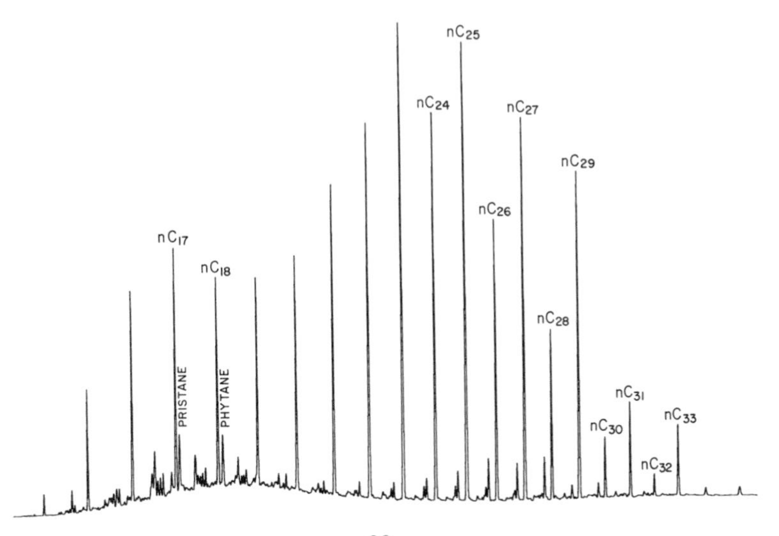

maturity until even and odd alkanes are present in equal amounts, i.e. an index of 1.0. Hence a high CPI, >1.1, means that an oil or extract is *immature. However, a low CPI does not necessarily mean that an oil or extract is *mature, because the alkanes may be derived from plankton and *bacteria, which do not necessarily show an odd predominance. Synsedimentary bacterial activity may reduce the odd numbered predominance. In highly reducing, evaporitic environments there may be an even predominance. *Biodegradation may cause an irregular alkane distribution, leading to an anomalous CPI.

Oils do not normally show a high CPI as they are derived from source rocks which are mature, most oils having a ratio of 1.0 ± 0.1. Greater deviations from 1.0 may indicate that oil has been expelled from the source rock much earlier than is usual; this is often encountered in source rocks where lithologies are other than shales. Source rock CPIs do not reach a value of 1.0 until mature to *late mature. This may indicate the difference between the maturity necessary for the generation of oil and that necessary for *expulsion of oil from a source rock. Different analysts use different carbon number ranges for calculating CPI. The full range of carbon numbers which have been included in these calculations is 20 to 34, but most analysts prefer to use a more restricted range. The most widely used is that published by Bray and Evans. It is:

$$\frac{1}{2}\left[\frac{(C_{25}+C_{27}+C_{29}+C_{31}+C_{33})}{(C_{24}+C_{26}+C_{28}+C_{30}+C_{32})}+\frac{(C_{25}+C_{27}+C_{29}+C_{31}+C_{33})}{(C_{26}+C_{28}+C_{30}+C_{32}+C_{34})}\right].$$

The effect of using different carbon-number ranges alters the precision with which maturity can be defined, i.e. 1.07 may be mature on one scale but immature on another. Errors in CPI determinations may arise where other compounds (such as *steranes and *triterpanes), co-elute with normal alkanes in immature oils and source rocks. Normal alkanes can be removed from a sample by using a *molecular sieve or *urea adduction then released from the molecular sieve or urea and the CPI measured without interference. *See also*: odd–even predominance. *Reference*: Bray, E.E. and Evans, E.D. (1961). Distribution of n-paraffins as a clue to the recognition of source beds. *Geochim. et Cosmochim. Acta*, **22**, 2–9.

carotenoids The most important tetraterpenoids. They are *irregular isoprenoids made up of eight *isoprene units (C_{40}). They are usually red to yellow pigments which occur in plants and *algae, mainly concentrated in the fruits and leaves of land plants. They are good environmental indicators, but rarely survive *diagenesis. Especially worthy of note is β-carotane $C_{40}H_{78}$ (see Fig., p. 41), which frequently occurs in extracts and oils derived from *lacustrine organic matter.

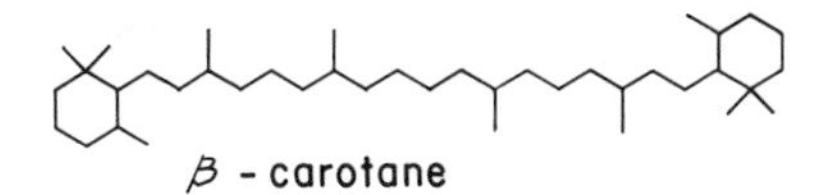

carrier rock *See* conduit.

catagenesis The principal *hydrocarbon generation zone. Elimination of hydrocarbon side-chains and alicyclics takes place, and *kerogen eventually reaches a maturity equivalent to anthracite *coal rank, with no further hydrocarbon generation. The equivalent *vitrinite reflectance range would be 0.5 to 2.0 per cent. Synonymous with katagenesis.

cellulose A water-insoluble *carbohydrate, synthesized by most higher land plants, e.g. wood (composed of cellulose and *lignin). It is classed as a polysaccharide (see Fig.) and it is degraded to water-soluble monosaccharides (sugars) in the water column and near-surface sediments.

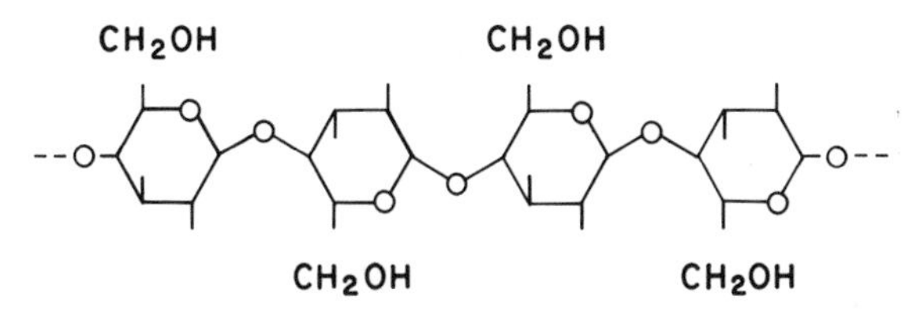

Cellulose

characteristic ions *See* diagnostic ions.

chemical kinetics The branch of physical chemistry which studies the quantitative relationships between velocity of a reaction, concentration of reagents, temperature of reaction, and composition of reaction medium. In a reaction, the *rate of a reaction is proportional to a function of the concentration of the reactants. This is written

$$\frac{\mathrm{d}c}{\mathrm{d}t} \propto f(c)$$

or

$$\frac{\mathrm{d}c}{\mathrm{d}t} = k(c)^n,$$

where c is the concentration, t is the time, k is the *rate constant, and n is the *order of the reaction. The rate constant varies with temperature, and this variation can be described using the *Arrhenius rate equation.

In *organic geochemistry, there are many direct applications of these kinetic laws but two are topical. The first is that the conversion of *kerogen to oil with increasing maturity may be estimated from the

Arrhenius rate equation if *activation energies, *A factors, and temperature history are known. This is known as *maturation modelling. The second is that if the concentration of reactants and products for a suitable set of reactions can be measured, the temperature history can be deduced. This is known as *inverse modelling.

The first reactions to be used were *amino acid isomerizations, which were found to be useful in fairly young sediments. More recently *steroid aromatization and isomerizations and *hopane isomerizations have been used for older sediments. A second approach has been to use *vitrinite reflectance to calculate *heatflow, although the kinetics of this transformation are more complex. *See also*: inverse modelling, maturation modelling. *Reference*: Abbott, G.D., Lewis, C.A., and Maxwell, J.R. (1985). The kinetics of specific organic reactions in the zone of catagenesis. *Phil. Trans. R. Soc. Lond.*, **A315**, 107–22.

chiral The name given to an asymmetric carbon at which *optical isomers may occur in an organic molecule. The carbon is referred to as a chiral centre.

chlorophyll The green pigment of plants which is necessary for *photosynthesis to occur. In sediments it breaks down into two distinctive fragments; the nucleus converts via porphins to *porphyrins, and the side chain converts via *phytol to *pristane and *phytane. There are many types of chlorophyll but the two main types are a and b, (see Fig.).

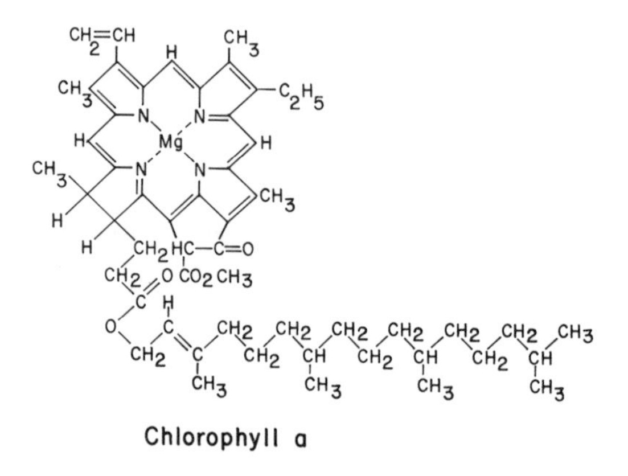

Chlorophyll a

cholestane C_{27} *sterane (see Fig., p. 43). *See also*: steranes.

chromatography A technique for separating mixtures of chemical compounds based on their physical and chemical properties, routinely used in *organic geochemistry. *Liquid chromatography is used to separate oils and *extracts into their bulk components, e.g. *saturate

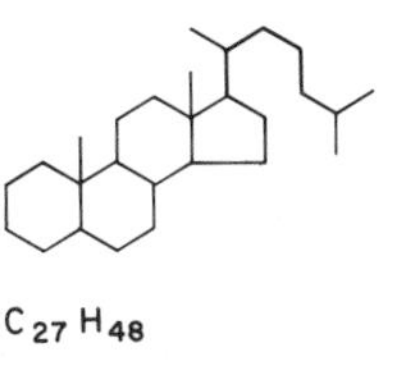

$C_{27}H_{48}$

Cholestane

and *aromatic fractions. *Gas chromatography is used to investigate the individual components of the saturate or aromatic fractions, sometimes prior to mass spectrometry. *See also*: column chromatography, gas chromatography, HPLC, TLC.

***cis–trans* isomerism** *See* geometrical isomerism.

clathration *See* adduction.

CNE Carbon Normalized Extract. *See* extract to organic carbon ratio.

coal A concentration of fossilized organic matter which is greater than 50 per cent of a rock by weight is called a coal. Such concentrations are called seams or beds. Coals can be subdivided into types based on their parent plant types. *Humic coals, which are the major coals, contain organic matter derived from woody debris, *cannel coals contain large amounts of spore and *resin material and *boghead coals contain algal-derived organic matter. They may also be divided on the basis of maturity into ranks. *See also*: coal rank.

coal gas Gas which is derived from the natural or artificial thermal breakdown of *coal is called coal gas. *See also*: natural gas.

coal rank The rank of coal is a measure of its degree of thermal alteration or maturity. Brown *coals and *lignites are described as low maturity coals; sub-bituminous, high, medium, and low volatile bituminous coals, semi-anthracites, anthracites, and meta-anthracites represent increasingly high levels of maturity (see Fig., p. 44). These ranks are related to calorific value, *vitrinite reflectance, volatile matter, and *FCC (fixed carbon contents) of coals. The *oil window is approximately equivalent to high and medium volatile bituminous coal ranks.

coaly An informal *kerogen type term which some analysts use to describe coal-like material, usually *inertinites. This type of kerogen has no oil, but may have some minor gas potential. In transmitted white light it may be dark brown or black, opaque or semi-opaque, and particulate. It may be synonymous with *semifusinite or *fusinite. The term coaly kerogen may also be used to describe mixtures of kerogen which have a coal-like maceral make-up in sediments, i.e. *vitrinite/

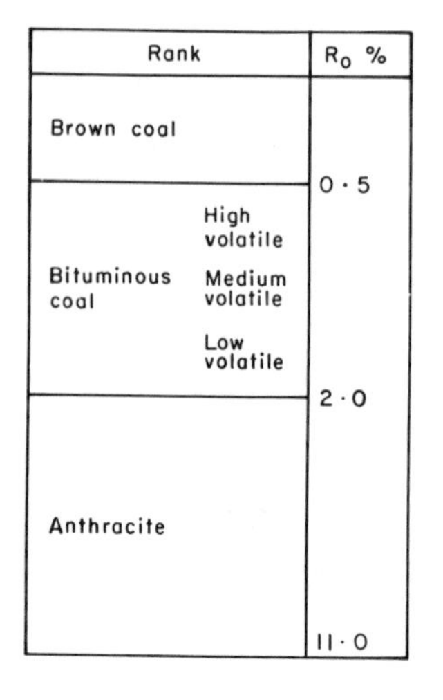

inertinite mixtures with some spores. It equates chemically with Type IV kerogen.

collinite Collinite is one of the *vitrinite group of *macerals. It is defined as vitrinite which is structureless in transmitted white light, and probably derives from colloidal *humic organic matter. It can be found as the filling of the cellular structures in *telinite. Synonymous with gelovitrinite. Chemically equivalent to *Type III kerogen. *See also*: telinite, vitrodetrinite.

column chromatography The type of chromatography commonly used to separate the bulk chemical components of *crude oils and *extracts. The separation is usually carried out on oils and extracts after *asphaltenes have been removed, or they precipitate in the column. The fractions are separated by virtue of their polarity, and the polarity of the solvents is increased during the separation of the different fractions.

The stationary phase is lightly packed powdered silica or alumina, or both, contained in a glass column. The silica and alumina have to be activated, or water removed, prior to analysis; the volume of silica is usually equal to, or greater than, the volume of alumina. The volume of the stationary phase is controlled by sample size. In the mobile phase, a series of solvents for each of the bulk components of oils, is passed through the column under gravity. Pentane, hexane, and heptane are the usual solvents for the *saturate fraction, *toluene or *benzene for the *aromatic fraction, and methanol for the *polar or *NSO fraction.

condensate A term used by petroleum engineers to describe *hydrocarbons which are in the gas phase under reservoir temperatures and pressures, but which form liquids and gas at the surface. Organic geochemists have adopted the term to imply a product which is dominantly gas and light oil, formed from sapropel-, spore-, or resin-rich organic matter at moderate to late maturity, or thermally decomposed oils. The character of a condensate may be illustrated by the relative amount of liquid compared to gas, i.e. barrels of condensate per cu.ft. of gas, or m^3 condensate per m^3 gas. *See also*: early mature condensate, retrograde condensate.

conduit A conduit is a porous, permeable horizon through which oils and gases migrate after leaving a source rock, i.e. during *secondary migration. Good conduits are laterally extensive and continuous, e.g. a beach barrier bar sand in a prograding sequence, or marine sheet sands. Fluvial sands only usually form good conduits in one direction. A fault may act as a conduit or a seal, depending on both its character and the lithology of the surrounding formations. Synonymous with carrier rock.

configurations *See* stereoisomers.

conversion index A measure of the progress of a chemical reaction. The conversion index is expressed as a percentage or decimal of 1. Zero per cent or 0.0, means that the reaction has not started, 50 per cent or 0.5, means that the reaction is half-way through, and 100 per cent or 1.0, means that the reaction is complete. In *organic geochemistry, the reaction is usually the conversion of *kerogen to oil. The term is specifically used in *maturation modelling based on the *Tissot and Espitalié Model. If a range of *activation energies is used to represent a series of simultaneous reactions, then the overall conversion index is the weighted index for each reaction.

coorongite A resistant deposit comprising the remains of *Botryococcus braunii*, found in Australia. It is believed to be the Recent equivalent of *boghead coal and *torbanite which are found throughout the sedimentary record. Synonymous with balkashite and boghead peat. *Reference*: Cane, R.F. (1969). Coorongite and the genesis of oil shale. *Geochim. et Cosmochim. Acta*, **33**, 257–65.

Corg *See* Total Organic Carbon.

CPI *See* Carbon Preference Index.

crude oil Unrefined oil, usually with gas in solution, as found in accumulations in the subsurface.

crude oil classification Crude oils may be compared using their bulk composition, usually on the basis of the relative amounts of *normal and *isoalkanes vs. *cyclic alkanes vs. *aromatic compounds and *NSO compounds. There are many different schemes; the classification shown (see Figs.), proposed by the Institut Français du Pétrole is a simple, logical classification applicable to most crude oils. *Reference*: Tissot,

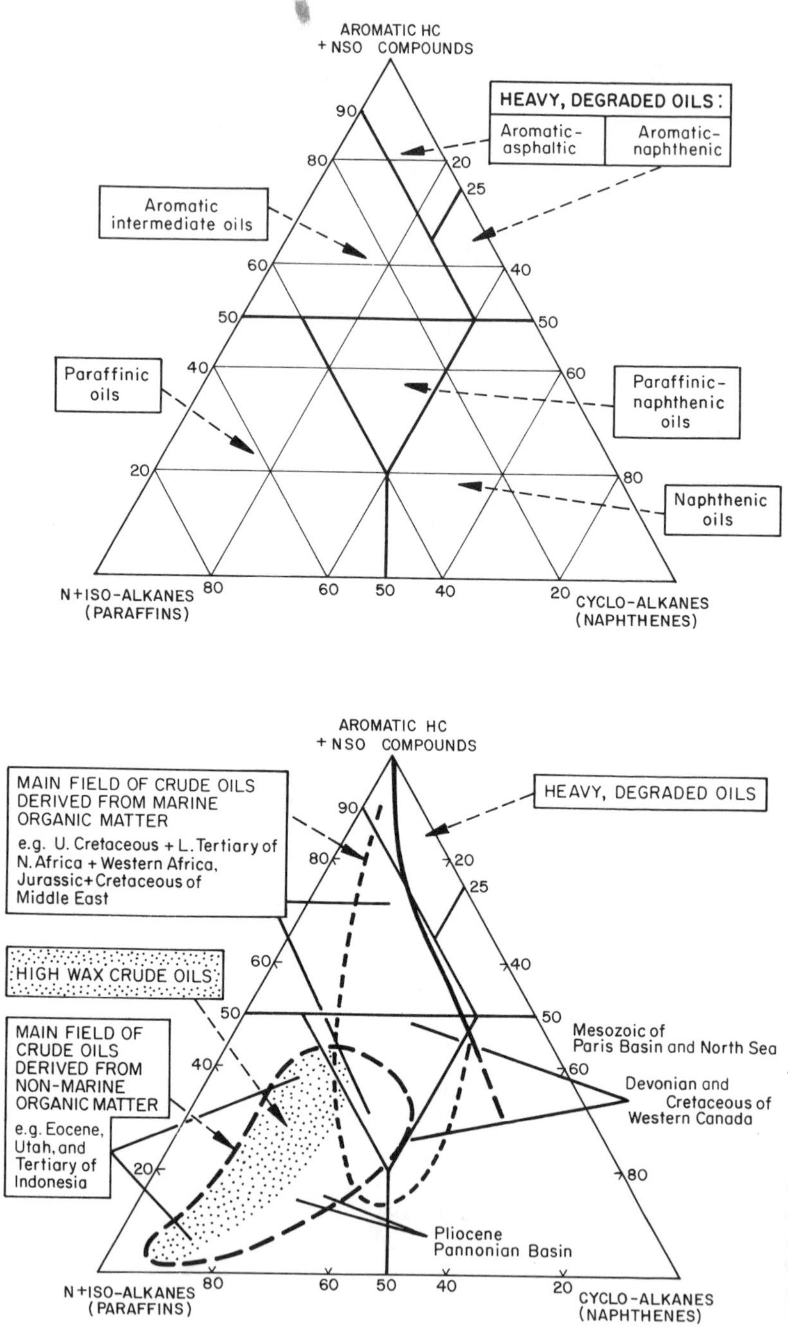
AROMATIC HC
+ NSO COMPOUNDS
HEAVY, DEGRADED OILS:
Aromatic-asphaltic
Aromatic-naphthenic
Aromatic intermediate oils
Paraffinic oils
Paraffinic-naphthenic oils
Naphthenic oils
N+ISO-ALKANES (PARAFFINS)
CYCLO-ALKANES (NAPHTHENES)
AROMATIC HC
+ NSO COMPOUNDS
MAIN FIELD OF CRUDE OILS DERIVED FROM MARINE ORGANIC MATTER
e.g. U. Cretaceous + L. Tertiary of N. Africa + Western Africa, Jurassic+Cretaceous of Middle East
HEAVY, DEGRADED OILS
HIGH WAX CRUDE OILS
MAIN FIELD OF CRUDE OILS DERIVED FROM NON-MARINE ORGANIC MATTER
e.g. Eocene, Utah, and Tertiary of Indonesia
Mesozoic of Paris Basin and North Sea
Devonian and Cretaceous of Western Canada
Pliocene Pannonian Basin
N+ISO-ALKANES (PARAFFINS)
CYCLO-ALKANES (NAPHTHENES)

B.P. and Welte, D.H. (1984). *Petroleum formation and occurrence* (2nd edn), 699 pp. Springer-Verlag, Berlin.

cutinite The *maceral name for the remains of the coverings of land plant leaves, shoots, roots, stems, and related tissue which are preserved in sediments. Sometimes the cellular structure may be preserved so that it is relatively easily identified. It is closely related to *suberinite, and occurs less frequently than *sporinite. It may fluoresce in UV light. It has light and waxy oil-generating potential and equates chemically with *Type II kerogen.

cuttings gas analysis During rotary drilling operations, cuttings from the borehole are brought to surface in the drilling mud and may be collected at the well head for gas analysis. To preserve the *hydrocarbon gases associated with the rock, the cuttings are collected in cans and sealed. Later, in the laboratory, samples of the gas can be removed from both the space above the mud, rock, and fluid sample (*headspace gas), and also from within the rock. The gas obtained from the rock is called the cuttings gas.

To release the gas, the rock is rapidly crushed in a grinder and the gases are vented to a gas chromatograph, where they are identified and the quantities of each gas present are noted. Normally gases up to pentane may be detected. This occluded gas is usually less contaminated than headspace gas; however it is depleted in *methane which tends to diffuse easily from the rock into the headspace in the can. This analysis can also be performed at the well site by the mud logging units. Results are used to aid both *kerogen type and maturity evaluations but care must be exercised as results may be altered not only by migrated gas, but also by contamination. *See also*: headspace gas analysis, light hydrocarbons.

cyanobacteria Originally used to describe blue-green *algae but now used synonymously with algae. *See also*: algae.

cyclic alkane A cyclic *carbon and *hydrogen compound which contains only carbon–carbon single bonds. Cyclic compounds forming rings of three or more carbon atoms are known in nature, but most

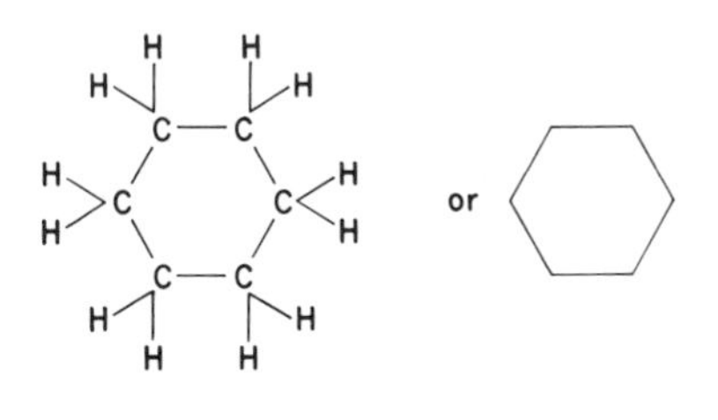

Cyclohexane

commonly the rings are five and six membered. The most important cyclic alkanes in petroleum geochemistry are the four and five ring compounds, *steranes and *triterpanes. Two and three ring compounds are also fairly common in sediments and oils. Synonymous with naphthenes and cycloparaffins. An example of a cyclic alkane is cyclohexane (see Fig., p. 47). *References*: Organic chemistry textbooks.

cycloalkanoporphyrin *See* DPEP.

cycloparaffin *See* cyclic alkane.

DAF Dry Ash Free. *See* FCC.

daughter ion The term which describes the decomposition product of the *parent (or *molecular) *ion, specifically during *metastable ion analysis, using a double focusing mass spectrometer. *See also*: metastable ions.

dead carbon The informal name given to *kerogen which has no oil and little gas potential even at high levels of *maturation. Equates chemically with *Type IV kerogen. Synonymous with *inertinite and *fusinite. *See also*: inertinite.

decarboxylation Elimination of CO_2 from an organic acid, $RCOOH \rightarrow RH$. These reactions occur mainly in shallow sediments at low temperatures, prior to the main zone of oil generation.

degraded vitrinite *Vitrinite which has become amorphous, or partially amorphous, due to the action of *bacteria on the parent *organic matter. This can be confused with *amorphous kerogen from a degraded phytoplanktonic source if white light alone is used in its examination. In UV light, degraded vitrinite does not fluoresce (except for a slight *fluorescence which may be present due to residual bacterial remains). The *hydrogen content may also be slightly elevated compared with undegraded vitrinite, although it is still chemically equivalent to *Type III kerogen. Synonymous with *amorphinite v. *See also*: amorphinite v.

demethylated hopanes Biologically (enzymically) transformed products of *hopanes found in biodegraded oils in the subsurface; the C-25 carbon, attached to the C-10 carbon, is removed from the hopane (see Fig., p. 49). (Note the loss of the methyl group from the A ring.) These compounds have never been produced during *in vitro* laboratory degradation experiments, and are not naturally produced by degradation of oils at the surface. The demethylation is not dependent on the structure of the whole hopane molecule, as demethylation of *tricyclic

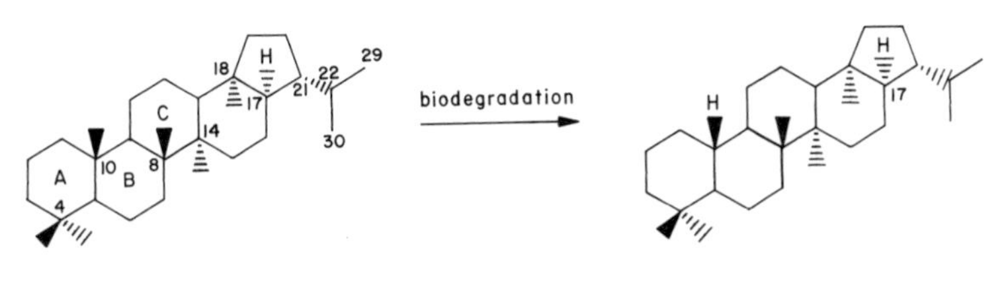

17α(H)-hopane 25-nor-17α(H)-hopane

terpanes occurs at the same carbon. They occcur as late-stage *biodegradation products, after removal of *normal and *branched alkanes. They are identified by *gas chromatography–mass spectrometry at *m/e* 177 (191 less a methyl). Synonymous with 25-norhopanes. *See also*: biodegradation. *Reference*: Volkman, J.K., Alexander, R., Kagi, R.I., and Woodhouse, G.W. (1983). Demethylated hopanes in crude oils and their applications in petroleum geochemistry. *Geochim. et Cosmochim. Acta*, **47**, 785–94.

detrovitrinite The maceral name for *vitrinite which occurs as detrital or attrital small fragments.

diacholestane The rearranged sterane equivalent of cholestane in regular steranes.

diagenesis Diagenesis of *organic matter in petroleum geochemistry refers to that stage of alteration where changes are dominantly due to biological, physical, and low-temperature chemical alteration, not severe thermal alteration processes. *Hydrocarbons are scarce, apart from *methane from *bacteria, and are mostly directly inherited from the parent plant and animal material. Chemical reactions are mainly classed as elimination reactions, with the breaking of heteroatomic bonds. Water and carbon dioxide are the main products. *Kerogen forms from the polymerization of material from the breakdown of plants. At the end of diagenesis all *decarboxylation reactions cease. It is equivalent to brown *coal to sub-bituminous *coal ranks, and equates to the *immature stage of *maturation with a *vitrinite reflectance of less than 0.5 per cent. Distinct from sedimentological usage. *See also*: catagenesis, metagenesis.

diagnostic ions Characteristic ions used to detect different classes of compounds in *gas chromatography–mass spectrometry analysis. Several examples of diagnostic ions for major compound classes in *organic geochemistry are *m/e*

85 normal alkanes,
113 botryococcane
123 + 272 diterpanes
183 acyclic isoprenoids

191 triterpanes
  + 123, 163 tri- and tetra-cyclic terpanes
  + 177 moretanes and demethylated hopanes
  + 205 hopanes and methyl hopanes
204 sesquiterpanes
217 steranes
  + 218 regular 14β(*H*) 17β(*H*)
  + 231 methyl steranes
  + 259 + 232 rearranged steranes
231 triaromatic steroids
245 methyl triaromatic steroids
253 monoaromatic steroids

Reference:Philp, R.P. (1985). *Fossil fuel biomarkers. Methods in geochemistry and geophysics*, Vol. 23. Elsevier, Amsterdam.

diasteranes *See* rearranged steranes.

diastereoisomers Isomeric compounds containing at least two *chiral centres, which differ in chemical and physical properties. They are *stereoisomers but not *enantiomers. *References*: Organic chemistry textbooks.

diterpenoids Common C_{20} constituents of higher plants comprising four *isoprene units (see Fig.). They are mainly cyclic, with two or three rings. *Phytol is an acyclic diterpenoid. They are particularly abundant in resins of coniferous plants, and their presence in oils is a diagnostic indication of sourcing from higher plants. Diterpanes are analysed by *gas chromatography–mass spectrometry of the *saturate fraction of oils and *extracts at *m/e* 272. *Reference*: Simonet, B.R.T., Grimalt, J.O., Wang, T.G., Cox, R.E., Hatcher, P.G., and Nissenbaum, A. (1986). Cyclic terpenoids of contemporary resinous plant detritus and of fossil woods, ambers and coals. *Org. Geochem.*, **10**, 877–89.

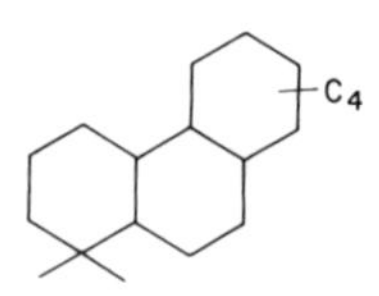

C_{20}-Diterpane

DPEP An abbreviation for de(*s*)oxophylloerythro(*a*)etioporphyrin. The transformation of DPEP to *etio porphyrins is a contentious *maturation parameter. They are detected by *gas chromatography–mass spectrometry at two *m/e* units lower than etio porphyrins and form an homologous series with *m/e* 478 + 14*n*. Synonymous with cycloalkanoporphyrin. *See also*: porphyrin.

dry gas Gases which contain over 95 per cent by volume of *methane are called dry gases. *Biogenic gas is classified as a dry gas as it usually contains more than 99 per cent methane. Highly mature gas from any source material may be dry; mature gas from *vitrinite and *semifusinite could also yield dry gas. *Ethane, *propane, and *butane are normally present in small quantities. *See also*: biogenic gas.

dysaerobic Loosely used to describe water conditions which are between *aerobic and *anaerobic. *Source rocks may be deposited under dysaerobic conditions but they are usually of fair to poor quality. Specifically used to describe water with a content of between 0.1 and 0.2 ml/l of dissolved oxygen. Synonymous with suboxic. *See also*: anaerobic, aerobic.

early mature The *maturation stage when *kerogen has generated some oil or gas within a *source rock, but before the main stage of hydrocarbon generation; between *immature and *peak mature. At this level of maturity, *primary migration may not have occurred except in carbonates or thinly interbedded source and *conduit or reservoir rocks, e.g. the Monterey Formation of California.

Early mature for *oil prone source rocks is taken as *spore colour index 3.5 to 5.0 (scale of 1 to 10), *Thermal Alteration Index (TAI), (Staplin) 2.2 to 2.3, *vitrinite reflectance 0.5 to 0.65 per cent. Early mature for *gas prone source rocks is TAI 2.5 to 2.6, vitrinite reflectance 0.7 to 1.3 per cent. Oils from early mature source rocks are called early mature crude oils. They are identified, although not exclusively, by low *API gravity and a low *saturate to aromatic ratio. *See* Table 1, summary section for maturation summary. *See also*: immature, peak mature, late mature, post mature.

early mature condensates Condensates which are believed to have formed at early *maturation levels from *source rocks which are rich in *kerogen derived from spores, pollen, and resin. *Reference*: Snowdon, L.R., and Powell, T.G. (1982). Immature oil and condensate. Modification of hydrocarbon generation model for terrestrial organic matter. *Am. Assoc. Petr. Geol. Bull.*, **66**, 775–88.

effective heating time The time, in millions of years, that a rock has spent within 15°C (or 27°F) of its maximum temperature in a continuously subsiding basin. It is used in *maturation modelling based on the method of *Hood *et al.*, and is calculated from the *geothermal gradient and *burial history. Long effective heating times are >25 m.y.; short values <10 m.y. Shortest values occur in Quaternary or Recent depocentres in continuously subsiding basins. *See also*: Hood *et al.* 1975.

effective source rock An effective source rock is one which has generated and expelled oil or gas. *See also*: potential source rock.

elemental analysis The analysis of the elemental composition of organic compounds is called elemental analysis. Elements C, H, O, S, N, and sometimes Fe, Ni, and V, are determined. *H/C and *O/C ratios are used to determine *kerogen type and maturity, H/C ratios are also used to determine the mode of formation of *asphaltenes in *crude oils. *Nickel/vanadium ratios can be an indicator of *source rock depositional environment. *See also*: asphaltenes, hydrogen index, oxygen index, Van Krevelen diagram.

eluent-eluant The mobile liquid phase used to separate chemical fractions of oil during *liquid chromatography is called the eluent. Pentane may be the eluent for the *saturate fraction, *toluene for the *aromatic fraction, and methanol for the *polar-NSO fraction. *See also*: liquid chromatography.

enantiomer The name given to the isomeric pairs in each case of *optical activity. Each enantiomer rotates the plane of plane polarized light in opposite directions. Examples of enantiomers are D and L or **R* (rectus) and **S* (sinister) forms of a *chiral molecule. They have identical physical and chemical properties but different spatial *configurations, which are non-superimposable mirror images. Interconversion of the enantiomers is called racemization. Synonymous with epimer, antimer, and enantiomorph. *See also*: optical isomer. *References*: Organic chemistry textbooks.

enantiomorph *See* enantiomer.

EOM Extractable Organic Matter. *See* extract.

epimer *See* enantiomer.

EPOC Extract Percentage Organic Carbon. *See* extract to organic carbon ratio.

ergostane C_{28} sterane. Synonymous with methyl cholestane. *See* regular steranes.

ESR Electron Spin Resonance. A technique which was used to measure the level of *maturation or *palaeotemperature, but results generally proved difficult to interpret, as they were also influenced by *kerogen type. An ESR signal results from an unpaired electron in the *kerogen. *Reference*: Pusey, W.C. (1973). The ESR kerogen method: a new technique of estimating the organic maturity of sedimentary rocks. *Petroleum Times*, 12th January. Whitehall Press, Kent, England.

ethane The *alkane containing two *carbon atoms, formula C_2H_6 (see Fig., p. 53). It is a gas at normal temperatures and pressures. Ethane is an

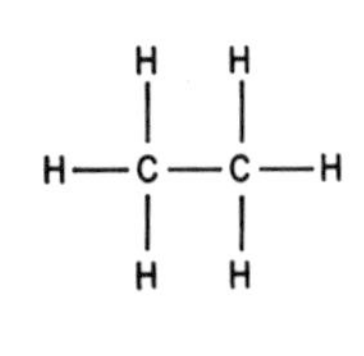

Ethane

abundant constituent of most *hydrocarbon accumulations except *biogenic gas; it is the second most abundant hydrocarbon after *methane. It is produced from both oil and gas prone *kerogen at all *maturation levels. It is a minor constituent of *dry gas.

ethylcholestane C_{29} sterane. Synonymous with sitostane.

etio porphyrin Porphyrins which may be thermally derived from *DPEP-type porphyrins, although it is also argued that etio porphyrins are independently formed. They are detected by *gas chromatography–mass spectrometry, and have *molecular ions two *m/e* units higher than DPEP, 480 + 14*n*. *See also*: porphyrin.

euphotic zone The upper part of a water column which is penetrated by sunlight, enabling *photosynthesis to proceed. The majority of organic matter production in the aqueous environment occurs in the euphotic zone. The depth of the zone varies with water temperature, salinity, current velocity, and sediment load but usually averages about 80m in open ocean. Synonymous with photic zone.

eutrophic Nutrient rich. Usually applied to water in lakes.

exinite A term used in the microscopic study of *kerogen which describes kerogen derived from the exines of spores and pollen, cuticle, resins, suberin, and phytoplankton. Exinite produces light oils and gas at the appropriate levels of *maturation. It is ubiquitous in all but oxidized sediments. Synonymous with *liptinite. It equates chemically with *Type I and *Type II kerogen.

expulsion The term used to describe the loss of oil and gas from a *source rock due to the increasing effects of temperature and pressure. The efficiency of expulsion of *hydrocarbons from most source rocks varies with lithology, organic carbon richness, maturity, and *kerogen type. It is believed that the expulsion of gas is more efficient than the expulsion of oil. Efficiencies in the range of 5 to 80 per cent seem to be appropriate from case studies. Synonymous with *primary migration. *See also*: primary migration, secondary migration.

exsudatinite A term used in the microscopic examination of *coal referring to the secondary *liptinite maceral generated during *maturation. It occurs in pore spaces and fractures, or as small veins. It is brown in transmitted white light and may fluoresce in UV light. It may be confused with *vitrinite, but has a lower reflectance, with values up to

1.1 per cent. Synonymous with secondary resinite, and sometimes used synonymously with bituminite, and bitumen although this is not correct usage. *Reference*: Teichmuller, M. (1977). Generation of petroleum-like substances in coal seams as seen under the microscope. In *Advances in Organic Chemistry 1975* (ed. R. Campos and G. Goni), pp. 379–407. Enadimsa, Madrid.

extract Oil and oil-like products which are removed from rock samples using *organic solvents are called extracts. Two extraction methods are commonly used, *ultrasonic and *soxhlet extraction. Ultrasonic extraction uses a cold solvent for short periods of time, commonly an hour or less. Soxhlet extraction is hot solvent refluxing, taking about 24 hours. Rocks are usually crushed or disaggregated. The techniques are used for *source rocks or for removing oil stains in reservoir rocks.

Extracts are examined for bulk compositional analysis, carbon isotopic composition and *biological markers. The amount of extract may also be used to determine the level of source rock *maturation and for estimates of the potential yield of a source rock in basin modelling. *Gas chromatography of the *saturate fraction of the extract can give information on the maturity of the sample or of its *kerogen type. Synonymous with total soluble extract, SOM, and bitumen. *See also*: bitumen, extract to organic carbon ratio, productivity.

extract to organic carbon ratio (per mil) The ratio of the amount of soluble organic matter or *bitumen extract in a *source rock related to the *total organic carbon, expressed as a per cent (%) or per mil (‰, mg/g). The ratio may be used to indicate the level of *maturation in an *oil prone source rock; *mature source rocks show values of 100 per mil to 200 per mil. The illustration (Fig.) shows the ratio typically varies with depth in oil prone source rocks. Values greater than 200 per mil usually indicate some form of contamination, or migrated oil stain. *Gas prone kerogen rarely gives values higher than 50 per mil at any

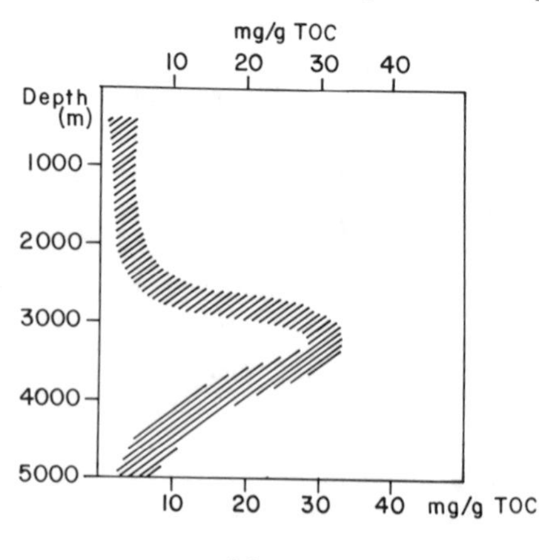

level of maturity. Synonymous with carbon normalized extract, total extract/organic carbon ratio, bitumen/TOC, extract to organic carbon per cent, EPOC.

fatty acids Fatty acids are carboxylic acids produced by plants and animals, usually with 4 to 40 carbon atoms. They may be either saturated or unsaturated, and usually have chains containing even numbers of carbon atoms. Only plants produce the unsaturated fatty acids. If they have more than one C=C double bond, they are called polyunsaturated fatty acids. They contain a polar carboxyl group with an affinity for water and a non-polar *hydrocarbon chain. They are part of the *lipid fraction of organic matter. They form *normal alkanes during *diagenesis in a sediment. They have the general formula $CH_3(CH_2CH_2)_nCO_2H$; when $n = 1$, the substance is butyric acid, and when $n = 8$ it is stearic acid.

FCC Fixed Carbon Content. A *coal rank scale based on the volatile matter and fixed carbon content of *coals or *vitrinite. An FCC of 0 per cent is 100 per cent volatile matter on a dry ash free basis (DAF); 40 per cent volatile is 60 per cent FCC, equivalent to a *vitrinite reflectance of 0.7 per cent; 25 per cent volatile is 75 per cent FCC and a vitrinite reflectance of 1.3 per cent. *See also*: volatile matter.

FID Flame Ionization Detector. The FID system is the normal *hydrocarbon detection system used in *gas chromatography. It operates on the basis that organic compounds produce ions when they burn in a *hydrogen flame, and cause a measurable current. *See also*: gas chromatography.

fingerprinting The name given to the method of correlating *oil to oil or *oil–source rock, usually by the comparison of the distribution of *biological markers present. *Isoprenoids, *steranes, and *triterpanes are most commonly used, although some *aromatic compounds are also useful. Fingerprints are obtained by *gas chromatography and *gas chromatography–mass spectrometry of the *saturate fraction of oils and *source rock extracts. The fingerprints may either be qualitative, by visual comparison of the patterns of various compounds present, or quantitative, by calculation of the relative and absolute amounts. Correlations between oils and source rocks are complicated by *maturation differences between rock and oil, and by migrational and biodegradation effects.

first-order reaction A reaction in which the *rate of reaction is proportional to the concentration of the reactant(s). If two or more compounds react, the reaction may be first order with respect to individual reactants but of a higher order overall. The decomposition of *kerogen is assumed to be a first-order reaction. *See also*: chemical kinetics. *References*: Physical chemistry textbooks.

Fixed Carbon Content *See* FCC.

Flame Ionization Detector *See* FID.

fluorescence The ability of a compound or substance to absorb UV light and to emit visible and some UV light is called fluorescence. It is caused by electrons being excited into a higher energy orbital by the absorbed energy, then falling back to their original states with the emission of energy in the form of light. The UV light emitted is of a shorter wavelength and lower energy than the absorbed UV light. Poor sample preparation may reduce sample fluorescence; fluorescence characteristics change with the time exposure to UV light, so determinations have to be carefully carried out. Another disadvantage is that *liptinite, which fluoresces, is not ubiquitous in sediments, thus limiting the number of samples suitable for analysis. *Aromatic compounds cause fluorescence in oils and *kerogen, although the phenomenon is poorly understood. Fluorescence is diagnostic of liptinitic or *oil prone kerogen.

Fluorescence colour is indicative of the level of maturity of kerogen and different *kerogen types have different characteristic fluorescence spectra, so that it may be possible to identify the source of *amorphous kerogen. Although presently not a routine analysis, it is possible to identify kerogen and assess its maturity from fluorescence spectra. Immature oil prone kerogen fluoresces yellow-green, and as the kerogen becomes more mature the colour changes through yellow, orange, and brown until it ceases at high maturity levels. *Sporinite (see Fig.), *alginite, and *cutinite are suitable for maturity measurements,

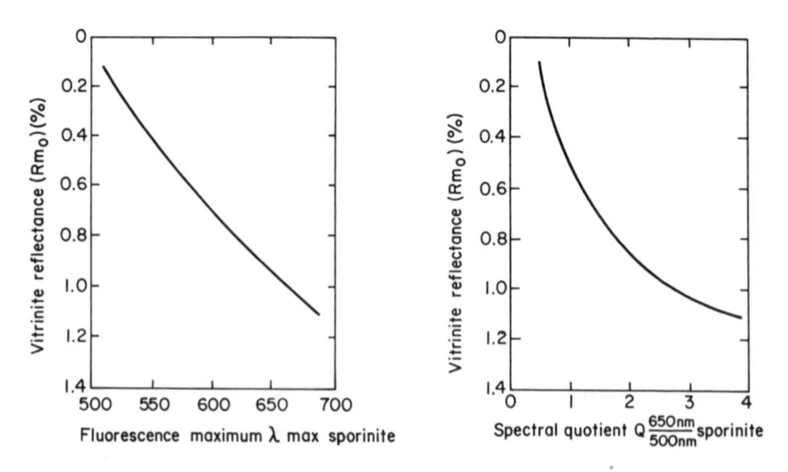

although not *resinite, as it can occur in several different forms, all of which may be present in a single sample. The two parameters which are the most reliable and reproducible are the spectral quotient (red/green), which is the ratio of the intensity of fluorescence at 650 nm and 500 nm for a particular kerogen type; and the maximum fluorescence wavelength, λ_{max} (see Fig., p. 56). *Torbanites have the characteristic known as negative fading, where the fluorescence intensity decreases after about 30 mins; with *tasmanites*, a type of alga, the fluorescence intensity increases with time of exposure. The fluorescence colour of *crude oils indicates their *API gravity. *See also*: API gravity. *Reference*: Khorasani, G.K. (1987). Novel development in fluorescence microscopy of complex mixtures: application in petroleum geochemistry. *Org. Geochem.*, **11**, 157–68.

fluorinite The *maceral derived from plant oils. It is recognized by strong *fluorescence in UV light, but is black in reflected white light, and transparent or translucent in transmitted light. It may be used synonymously with *resinite.

fractionation *See* stable carbon isotopes, sulphur isotopes.

frequency factor *See* A factor.

fulvic acids Those organic acids found in soils and shallow sediments which can be extracted using sodium hydroxide and sodium pyrophosphate. Fulvic acids are the acid-soluble portion of the extract (see Fig.). They derive from the breakdown of plant material during *diagenesis.

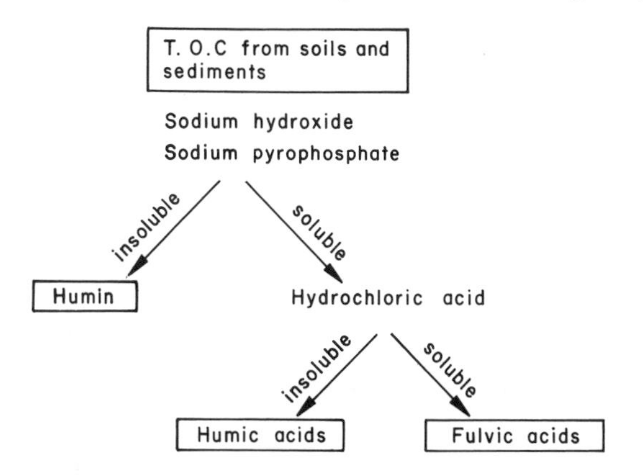

fusinite The coal petrological term used to describe one of the *inertinite group of *macerals in the microscopic study of *kerogen. It is the richest in *carbon, and occurs as black or opaque angular fragments in transmitted white light; it is highly reflecting, white to grey-white in reflected light. Fusinite may be formed in two ways; by high temperature oxidation during fires (*pyrofusinite), where cell lumens are often preserved; or low-temperature oxidation in a fluctuating water table (oxyfusinite and degradofusinite). The roundness and size of the particles reflects the energy levels in the depositional environment of the sediment, large angular particles being deposited in low energy environments and small well-sorted particles in high energy environments. Synonymous with *dead carbon and *coaly. Equates chemically with *Type IV kerogen.

gammacerane A pentacyclic *triterpane made up of only six-membered rings, formula $C_{30}H_{52}$ (see Fig.). It is a facies-controlled triterpane, occurring predominantly in *non-marine environments. Its abundance compared with $17\alpha(H)$, $21\beta(H)$-hopane(C_{30}) expressed as a per cent, is called the Gammacerane Index. High amounts of gammacerane may indicate hypersaline *lacustrine environments. It is resistant to *biodegradation so its abundance may be enhanced in biodegraded oils. There is evidence that it migrates less efficiently than the other triterpanes so its abundance may be enhanced in source rocks relative to oils. Its abundance is often linked with the occurrence of 4-*methyl steranes. It is recognized by *gas chromatography–mass spectrometry of the *saturate fraction of oils and *extracts with *m/e* 191, the absence of a *m/e* 369 fragment and relative retention time (although this alone is not diagnostic, as other triterpanes have a similar retention time).

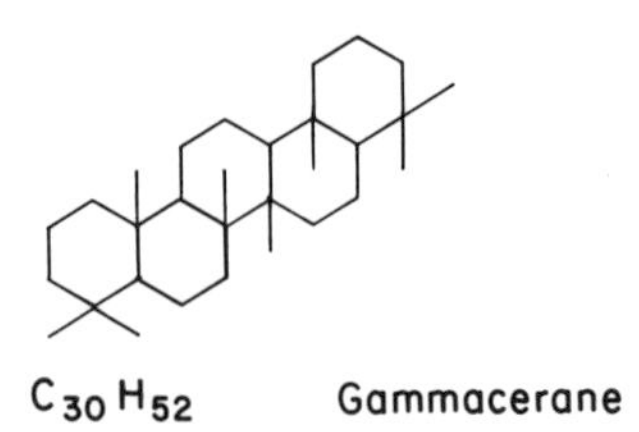

$C_{30}H_{52}$ Gammacerane

gas chromatography The separation of mixtures of compounds by partition between a mobile gas phase and stationary liquid phase is called gas chromatography (GC or GLC). Two types of gas chromatography are used routinely in organic geochemistry. Packed column gas chromatography is used to analyse *hydrocarbon gases from mud, cuttings gas and headspace gas, or *gasoline range hydrocarbons from oils and *condensates. Capillary column chromatography is used to separate components of oils and extracts, especially the *saturate fraction.

In both types of gas chromatography the sample is introduced by a calibrated syringe through a self-sealing septum and carried by a gas (such as *hydrogen or helium) over the stationary phase. Separation occurs by differential partition of sample components between the gas

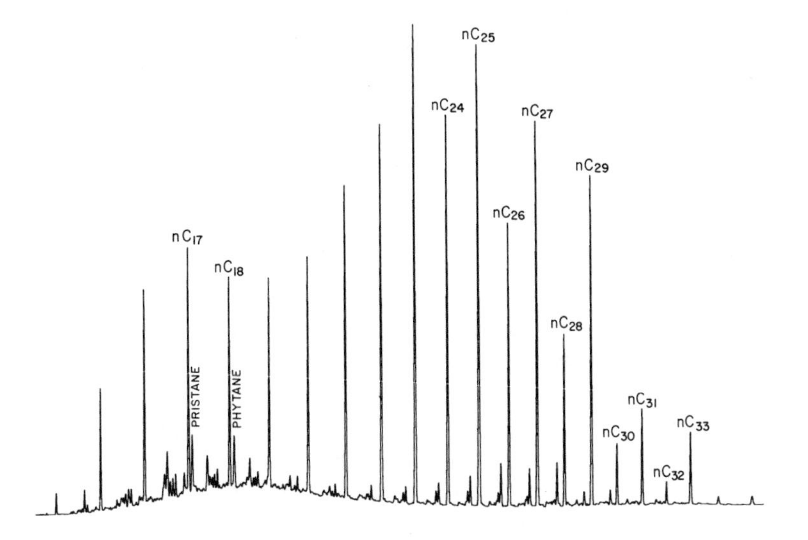

and liquid phases. Components exit the distal end of the column at times dependent on their retention in the liquid phase. These times are called retention times, and they depend on many factors including the temperature of the column. In packed column chromatography, the column is usually of metal and filled with a powder coated with a high molecular weight *alkane. It is short (25 cm) and of wide bore (1–2 cm) and held isothermally, the temperature depending on the compounds to be analysed. In capillary chromatography, the column is made of glass or fused silica, and is of narrow bore (<1 mm) and of lengths varying from 10 to 60 m. The stationary phase is usually bonded to the inside of the capillary column.

Samples may be liquid at room temperature, so that it is necesary to vaporize them in a heated injector port prior to introduction onto the column. As there is usually a wide range of components present in the sample, some of the retention times would be extremely long if the column were held isothermally. To ensure that analysis is complete in a realistic time, the column is heated at a rate dependent on the operator's needs and sample requirements. A typical heating rate would be 4°C per minute from a starting temperature of 50 to 200°C to a final temperature of 350°C so that an analysis takes about an hour.

As the compounds leave the column they are detected by a system such as a *flame ionization (FID) or *thermal conductivity detector. The detector response is recorded on a chart, the compounds being detected as a series of peaks, the area under each being proportional to the amount of compound present. Most gas chromatographs are connected to laboratory computers and the data is stored on discs.

Good resolution refers to the ability of the system to separate components with close retention times. In most oils and *extracts, *normal alkanes appear as dominant peaks. *Pristane and *phytane are also usually present, and occur as distinctive doublets with normal alkanes with 17 and 18 carbon atoms (see Fig., p. 59), so that the experienced operator can soon identify the components present (compounds may also be identified by co-injection of the sample with a standard). Ratios such as *pristane/phytane and *Carbon Preference Index (CPI), can be calculated from the traces. The input to the GC may be modified to allow the analysis of *kerogen *pyrolysis products. It is then called pyrolysis-GC or PY-GC. The gas chromatograph may be used as the input system for further analysis, such as mass spectrometry. *See also*: gas chromatography–mass spectrometry. *Reference*: Douglas, A.G. (1969). In *Organic Geochemistry* (ed. G. Eglinton and M.T.J. Murphy), 828 pp. Springer-Verlag, Berlin.

gas chromatography–mass spectrometry, (GC–MS) The name given to the tandem system of a *gas chromatograph linked to a mass spectrometer. This equipment is the most powerful analytical technique in modern *organic geochemistry. The technique is used for compound separation and identification, especially higher molecular weight *cyclic alkanes, *aromatic steroids, and *porphyrins. The compounds elute through the gas chromatographic column according to their retention times. As compounds leave the column, they enter the mass spectrometer where they are bombarded by a high energy beam of ions from an ion source. This causes the compounds to break into patterns of fragments determined by their molecular structure. Each GC peak contains many individual molecules of the same compound, and not all of them break in exactly the same way, so a spectrum of fragments is obtained. These vary from the complete molecule, known as the *molecular ion, to very small non-diagnostic fragments. They are detected and separated according to their m/e or m/z ratio, which is their mass to charge ratio. Different types of compounds have preferred fragmentation patterns, leading to *diagnostic ions. The *triterpanes fragment with a dominant m/e of 191 and *steranes m/e 217. At any one time the relative intensity of fragments of each m/e number can be recorded (see Fig., pp. 61–2). The frequency of acquisition of these records, called scans, is determined by the operator. A typical number of scans for a sample would be 3000 in a period of one to one and a half hours. This gives a large amount of information and most GC–MS systems have their own dedicated computer.

The data may be viewed in two ways. For compound identification, all the fragments at one time are inspected; the display is known as a mass spectrum (see Fig., p. 63). Most data systems are equipped with a library against which the fragment spectrum can be compared and an identification made. Another way of viewing the data is to examine the

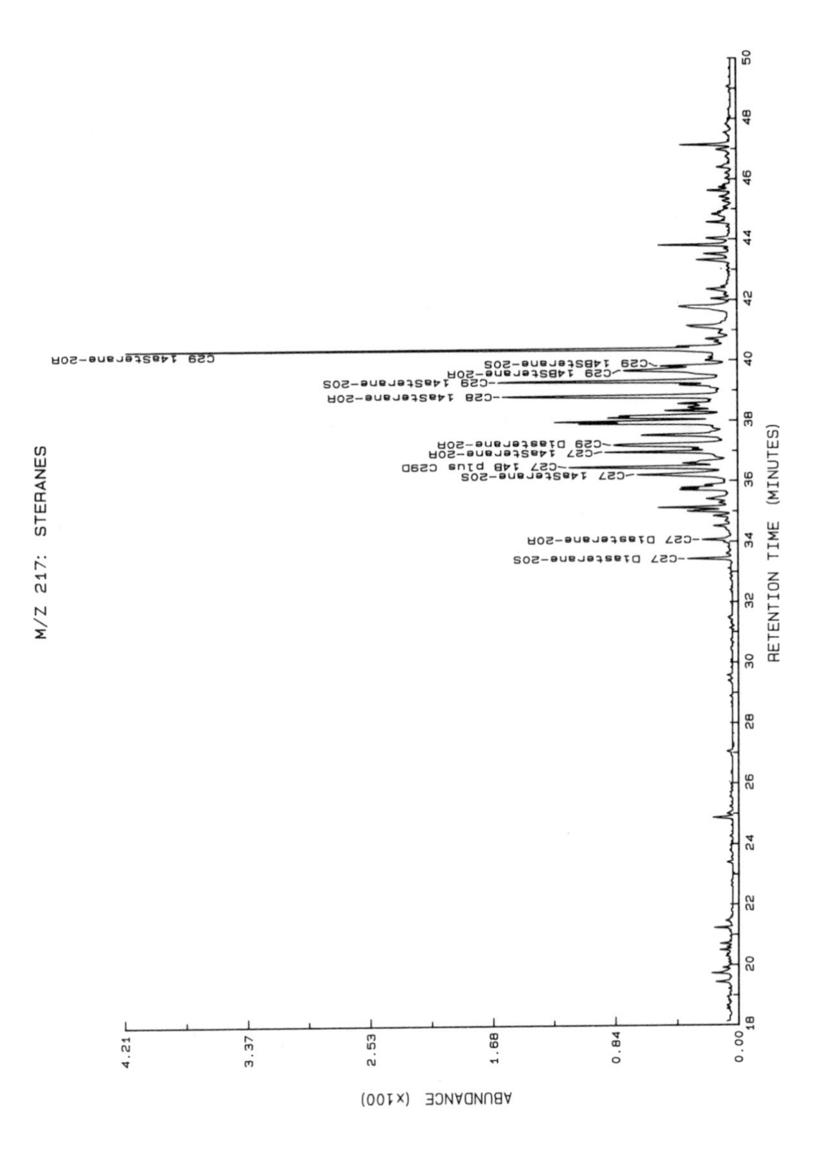
M/Z 217: STERANES
C27 Diasterane-20S
C27 Diasterane-20R
C27 14aSterane-20S
C27 14B plus C29D
C27 14aSterane-20R
C29 Diasterane-20R
C28 14aSterane-20R
C29 14aSterane-20S
C29 14BSterane-20R
C29 14BSterane-20S
C29 14aSterane-20R
ABUNDANCE (x100)
0.00
0.84
1.68
2.53
3.37
4.21
RETENTION TIME (MINUTES)
18
20
22
24
26
28
30
32
34
36
38
40
42
44
46
48
50

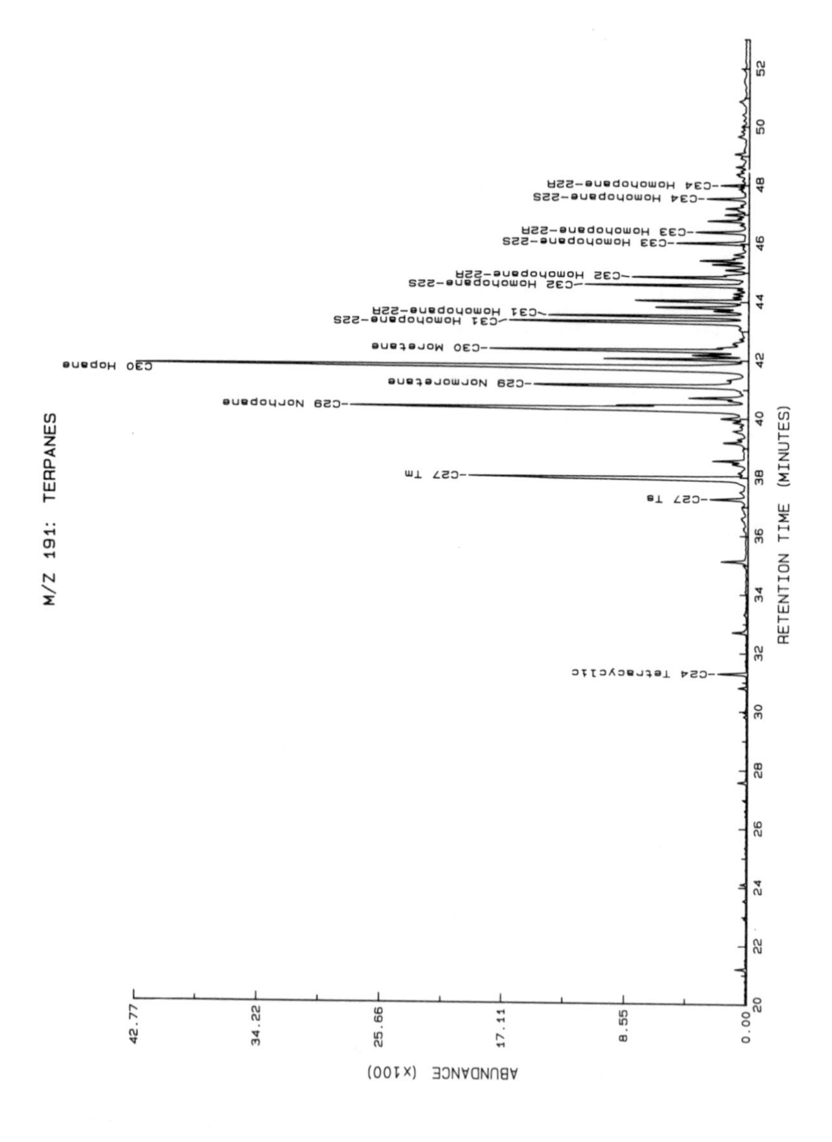
M/Z 191: TERPANES
RETENTION TIME (MINUTES)
ABUNDANCE (x100)
42.77
34.22
25.66
17.11
8.55
0.00
20
22
24
26
28
30
32
34
36
38
40
42
44
46
48
50
52
C24 Tetracyclic
C27 Ts
C27 Tm
C29 Norhopane
C29 Normoretane
C30 Hopane
C30 Moretane
C31 Homohopane-22S
C31 Homohopane-22R
C32 Homohopane-22S
C32 Homohopane-22R
C33 Homohopane-22S
C33 Homohopane-22R
C34 Homohopane-22S
C34 Homohopane-22R

intensity of the record of one particular *m/e* number with time; for example 217 for steranes. This produces a trace which looks like that of a GC (see Fig., pp. 61–2). This method is routinely used in applications to petroleum geochemistry and is known as Single Ion Monitoring or *SIM. Several ions may be run to improve compound identification; this is called Multiple Ion Detection or *MID. Other data manipulations are possible depending on the sophistication of the GC–MS system. A different type of mass spectrometer is used for the measurement of the abundance of different isotopes of the same element. *See also*: diagnostic ions, molecular ions. *Reference*: Hill, H.C. and Loudon, A.G. (1972). *Introduction to mass spectrometry* (2nd edn), 116 pp. Heyden and Sons, London.

gas–gas correlation The correlation of *natural gases to each other, usually by means of bulk compositional analysis and *stable carbon isotope ratios, and isotope ratios of other elements present. A gas may not usually be correlated to a specific *source rock but rather an environment of deposition, *kerogen type, and *maturation level.

gas hydrate Complexes of *methane and water found in deep water settings. They may be identified as bottom-simulating reflectors on seismic sections. They are stable under a restricted range of temperature and pressure conditions. They are usually formed from methane generated biogenically in near-surface sediments, but could also be formed from diffusing *thermogenic gas. *Reference*: Kvenvolden, K., Claypool, G.E., Threlkeld, C.N., and Sloan, E.D. (1984). Geochemistry of a naturally occurring, massive marine gas hydrate. *Org. Geochem.*, **6**, 703–13.

gasoline range hydrocarbons *Alkanes and *aromatic hydrocarbons of carbon numbers up to C_7 in oils and condensates are known as

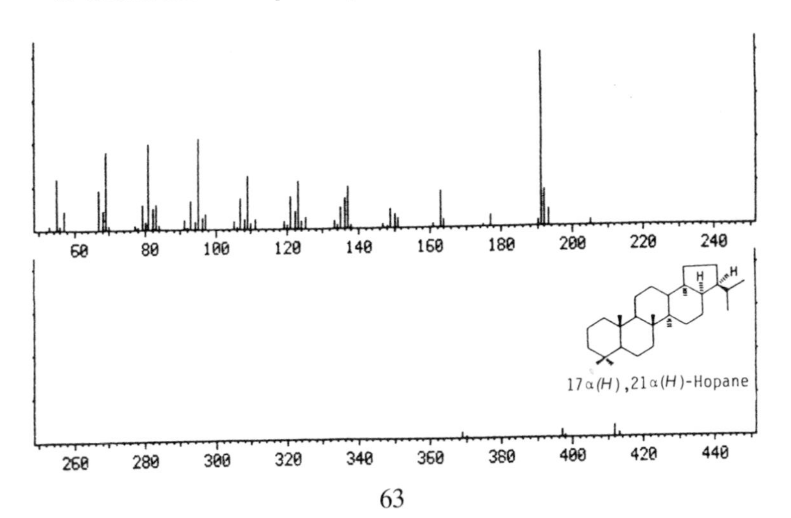

gasoline range hydrocarbons. This fraction contains many *structural isomers of alkanes, whose relative abundances can be used to derive source quality and maturity parameters. As this fraction is volatile at ambient temperatures, it can be partially or totally lost by poor sample handling. *See also*: butane, heptane value, isoheptane value.

gas prone A sediment or *coal containing *kerogen which at optimum maturity will only produce *hydrocarbon gases with minor associated liquids is called gas prone. This kerogen is dominantly terrestrially derived, e.g. *vitrinite, and equates chemically with *Type III kerogen.

gas wetness The amount of *hydrocarbon gases (other than *methane) in a gas sample, expressed as a percentage of the whole sample, by volume, is known as its wetness,

$$\frac{C_2 + C_3 + C_4}{C_1 + C_2 + C_3 + C_4} \times 100.$$

A gas is defined as wet if it contains more than 5 per cent C_2 to C_4. Wetness may indicate the origin of the gas, e.g. 0.01 per cent wet is *biogenic gas. *Oil prone *source rocks which are mature emit gases which can be up to 90 per cent wet.

GC *See* gas chromatography.

GC–MS *See* gas chromatography–mass spectrometry

GC–MS–MS Gas chromatography–mass spectrometry–mass spectrometry. Analysis by a double focusing mass spectrometer. The decay of *parent ions to *daughter ions can be observed, so that the source of *diagnostic ions can be discerned. *See also*: metastable ions.

gelovitrinite *See* collinite.

geochemical fossils *See* biological markers.

geometrical isomerism A form of *stereoisomerism commonly found in compounds containing rings and C=C bonds. *Isomers are known as *trans* and *cis* forms; in complex molecules they are represented at a carbon atom by the *α and β configuration. The two isomers may have different physical properties and chemical activities. Synonymous with *Z* and *E* isomerism, and *cis–trans* isomerism. An example of two geometrical isomers is *cis*- and *trans*-crotonic acid (see Fig.). *References*: Organic chemistry textbooks.

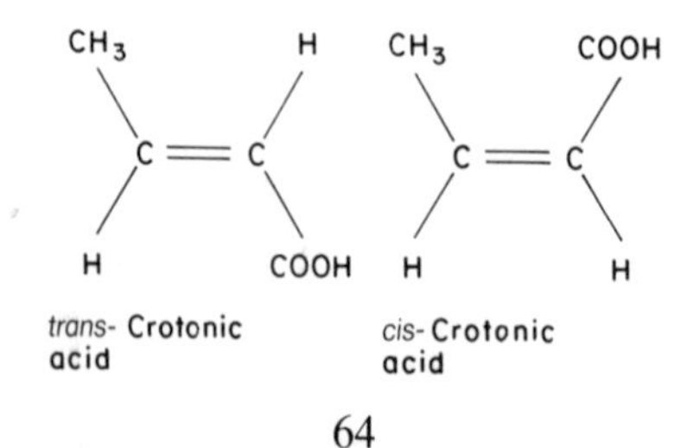

geothermal gradient, (ΔT) The measure of the rate of change of temperature with depth in the earth's crust is known as the geothermal gradient. It is measured as °C/km or °F/100 ft., usually from borehole logging data. Gradients are usually calculated from the surface to a specified depth, and derived from a series of values for different depth intervals at a single location, which allows for the cooling effect of drilling fluids on the formations. Another source of reliable temperatures are bottom hole flowing temperatures (BHFT) for oil. In the absence of heat redistribution by fluid flow (convection), the geothermal gradient depends on the *heatflow and *thermal conductivity (conduction). However, examination of data for many areas shows that there is rarely a total absence of convection. Worldwide average geothermal gradients are from 1.3 to 2.2 °F/100 ft. (or 24 to 41 °C/km), with extremes of 0.3 to 4.9 °F/100 ft. or 5 to 90 °C/km. *See also*: thermal conductivity, heatflow. *Reference*: Fertl, W.H. and Wichmann, P.A. (1977). How to determine static BHT from well log data. *World Oil*, **184**, 105–6.

gilsonite *See* solidified bitumen.

GPI Gas Production Index. A parameter derived from the *Oil Show Analyzer, S_0 peak.

grahamite *See* solidified bitumen

gravity *See* API gravity, specific gravity.

H/C ratio The atomic ratio of *hydrogen to *carbon determined by *elemental analysis, usually of *kerogen or *asphaltenes. The H/C ratio may be plotted with *O/C ratios on a *Van Krevelen diagram to show *kerogen type and maturity differences. H/C elemental ratios are less frequently determined than the related *hydrogen index from *Rock Eval pyrolysis (see Fig., p. 66). A high H/C ratio, of >1.5, indicates *oil prone organic matter when *immature. H/C ratios may indicate the origin of the asphaltenes in oils. Those which are thermally derived have an H/C ratio of >0.53 and those which are derived from gas solution have an H/C ratio <0.53. *See also*: asphaltenes, hydrogen index, Van Krevelen diagram. *Reference*: Rogers, M.A., McAlary, J.D., and Bailey, N.J.L. (1974). Significance of reservoir bitumens to thermal maturation studies, Western Canada Basin. *Am. Assoc. Petr. Geol. Bull.* **58**, 1806–24.

headspace gas analysis The analysis by *gas chromatography of the *light hydrocarbons which collect in the headspace above caned

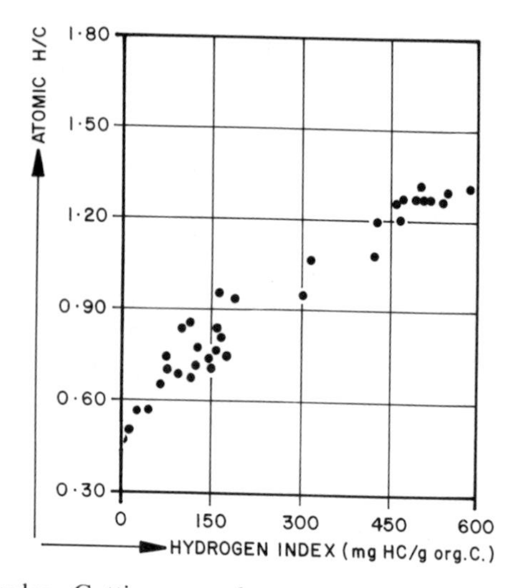

cuttings samples. Cuttings samples are collected by the mud logger during drilling operations and placed with bacteriocide in approximately one litre cans and sealed. Gas diffuses from the rock and drilling mud into the headspace. Samples are taken by syringe through a septum inserted in the wall of the can above the gas/liquid interface. Hydrocarbons from *methane to *butane are usually detected. The abundance and composition of the gas can be used to indicate the presence and *maturation of *source rocks, and the presence of reservoired hydrocarbons. *See also*: cuttings gas analysis.

heatflow, (Q) The crustal heatflow is the amount of heat energy leaving the surface of the earth per unit area and time. Apart from heat from the centre of the earth, heat is also produced by the decay of radioactive minerals. This is a significant heat source in igneous rocks, but less so in sedimentary rocks. One heatflow unit (HFU) is equal to 10^{-6} cal cm^{-2} s^{-1} (cgs units) or 0.42×10^{-5} Wm^{-2} (mks units). The relationship of heatflow to *thermal conductivity (k) and *geothermal gradient (ΔT) is:

$$Q = k \times \Delta T.$$

In most sedimentary basins, the heatflow is not constant with time but varies, according to the mechanism of basin formation. Changes in heatflow can be modelled to give estimates of temperature changes with time, providing the thermal conductivity of the rocks is known. These models are used in *maturation modelling. Heatflow due purely to conduction is constant with depth. If fluid flow causes convection, then heatflow ceases to be constant with depth. A rough worldwide average for heatflow is 1.5×10^{-6} cal cm^{-2} s^{-1} or 0.63×10^{-5} Wm^{-2}. *See also*: maturation modelling, rift basin models, thermal conductivity.

heavy (isotopes) Containing relatively more of a higher atomic weight isotope than a standard or another sample, e.g. more ^{13}C relative to ^{12}C. Heavy isotopic ratios are produced by different processes, e.g. *photosynthesis produces isotopically heavier organic carbon than the ratio in atmospheric CO_2, and *thermogenic methane is isotopically heavier than *biogenic methane. *See also*: stable carbon isotopes.

heavy oil A heavy oil is one of *API gravity less than 25°. Low API gravity is typical of oils which are from *immature *source rocks, or those which have been biodegraded or water-washed. *See also*: tar mat.

heptane value Heptane is an *alkane containing seven *carbon atoms. It has several *structural isomers whose relative abundance can be determined from *gasoline range hydrocarbon analysis. These data have been used to calculate a parameter known as the heptane value, which provides maturity and source information for oils and *condensates and *source rocks. The index is calculated as follows:

$$100 \times \frac{22}{15 \text{ to } 22 + 24}$$

where 15 is cyclohexane, 16 is 2-methylhexane, 17 is 1,1-dimethylcyclopentane, 18 is 3-methylhexane, 19 is *cis*-1,3-dimethylcyclopentane, 20 is *trans*-1,3-dimethylcyclopentane, 21 is *trans*-1,2-dimethylcyclopentane, 22 is n-heptane, and 24 is methylcyclohexane. The figure shows a plot of heptane index against maturity, represented by *vitrinite reflectance. An index of 18 to 22 indicates a low maturity oil, 22 to 30 a mature oil, 30 to 60 a super-mature oil, 0 to 18 *biodegradation. The index should be used cautiously as it is measured on compounds which can be lost by poor sample handling. *See also*: isoheptane value. *Reference*: Thompson, K.F.M. (1983). Classification and thermal history of petroleum based on light hydrocarbons. *Geochim. et Cosmochim. Acta*, **47**, 303–16.

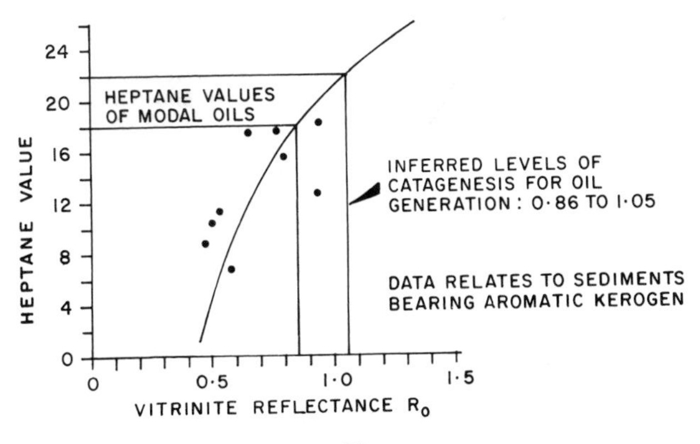

herbaceous An informal *kerogen typing term which includes all membranous plant material inclusive of cuticle, spores, and pollen. It may have both oil and gas generating potential. It equates chemically with *Type II kerogen.

hetero compounds Organic compounds which contain elements other than *carbon and *hydrogen, e.g. *sulphur, *oxygen, and *nitrogen. The term has also been applied to the fraction of oils and sediment extracts which elutes with methanol in *liquid chromatography. Synonymous with *polar fraction, *NSO, and *resins, depending on the solvent mixture used.

homohopanes C_{31} upwards *hopanes, which have a *chiral centre at the C-22 carbon. They are distinctive as *R* and *S* doublets during *gas chromatography–mass spectrometry analysis on a display of *m*/*e* 191. The ratio of *R* and *S* *isomers is used as a *maturation parameter. *See* Table 1, summary section, for maturation summary.

Hood *et al.* The method of calculating maturity published by A. Hood, C.C.M. Gutjahr, and R.L. Heacock in 1975. The method is based on an approximation to the *Arrhenius rate equation. An *effective heating time (T_{eff}), defined as the time taken to bring a sediment to within 15°C of its maximum temperature, is calculated from *burial history and *geothermal gradient data. It is used to compensate for the effects of both temperature and time on *maturation of *kerogen. The T_{eff} is then used graphically to derive the (higher) temperatures required to achieve various maturity levels expressed as *LOM. *Reference*: Hood, A., Gutjahr, C.C.M., and Heacock, R.L. (1975). Organic metamorphism and the generation of petroleum. *Am. Assoc. Petr. Geol. Bull.*, **59**, 986–96.

hopanes C_{27} to C_{35} pentacyclic *alkanes which dominate the *triterpanes found in sediments and *crude oils. They derive from *bacteria and are used for *oil to oil and *oil–source rock correlations. In their parent compound and *immature sediments they have 17β(*H*), 21β(*H*) stereochemistry, which changes to 17α(*H*), 21β(*H*) with increasing maturity (see Fig.). The higher carbon number hopanes (C_{31}+) have a *chiral centre at C-22, with *R* and *S* isomers, whose relative abundance

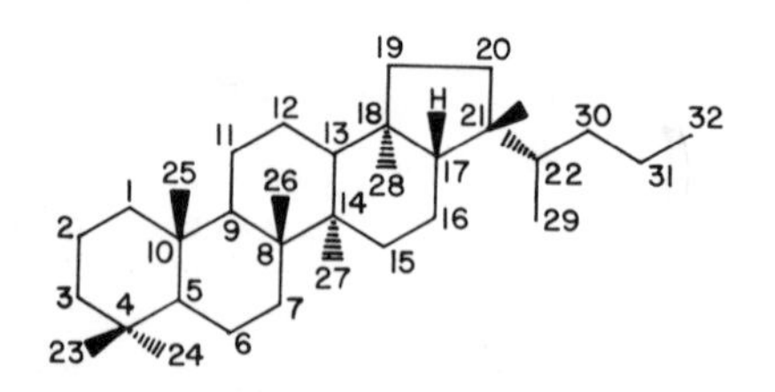

Hopane / Moretane

can be used as a maturation parameter. At equilibrium, the *S* isomer is slightly more abundant than the *R* isomer. The compounds with 17β(*H*), 21α(*H*) stereochemistry are called *moretanes, and their abundance relative to the hopanes can also be a *maturation parameter. They are analysed by *gas chromatography–mass spectrometry of the *saturate fraction of sediment *extracts and oils and their *diagnostic ion is *m*/*e* 191, and are generally characterized by 191> 177. *See also*: demethylated hopanes, moretanes, triterpanes. *Reference*: Ourisson, G., Albrecht, P., and Rohmer, M. (1979). The hopanoids: palaeochemistry and biochemistry of a group of natural products. *Pure and Appl. Chem.*, **51**, 709–29.

hopanoids The group term for different classes of compound containing the *hopane skeleton, e.g. hopenes, hopanols.

HPLC High Performance Liquid Chromatography is a technique which separates oils and *extracts into fractions, as in *liquid chromatography. Analyses take less time than conventional liquid chromatography, and can be fully automated. The technique is also more versatile, as it can allow more detailed separation of fractions, (for example *aromatic compounds by the number of aromatic rings), and can be used to separate *porphyrins from the *polar fraction of oils and sediment *extracts.

humic A general term for organic matter which is derived from vascular land plants. This includes *vitrinite, *huminite, *humin, *humic acids, and *fulvic acids. *Humic material is dominantly *gas prone and equates chemically to *Type III kerogen.

humic acids High molecular weight organic acids which can be extracted from soils and near-surface sediments by sodium hydroxide and sodium pyrophosphate (see Fig.). They are polymeric compounds which are derived from the breakdown of plant and microbial material

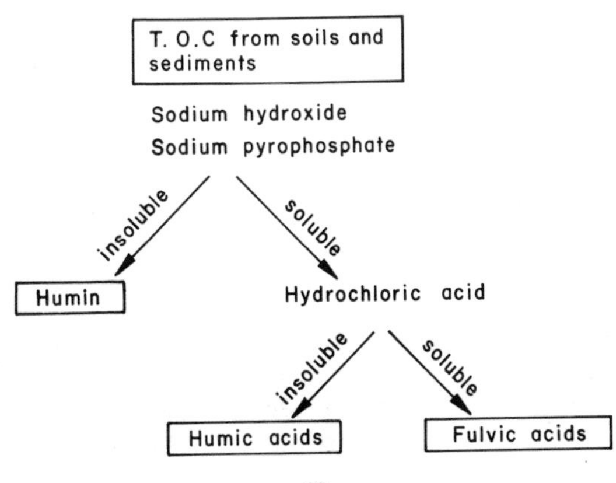

during *diagenesis under *aerobic conditions. Humic acids are the acid-insoluble portion of the *extract (*fulvic acids are the acid-soluble portion of the extract). *See also*: fulvic acids.

humin High molecular weight organic matter in soils and sediments (derived from plant and microbial sources) which is insoluble in sodium hydroxide and sodium pyrophosphate is called humin. It is a high molecular weight polymeric compound and is the precursor to *humic kerogen (see Fig., p. 69).

huminite The brown *coal maceral equivalent of *vitrinite in black coals. It is derived from the lignified tissue of higher land plants. It equates chemically to *Type III kerogen.

hydrate *See* gas hydrate.

hydrocarbon Organic molecules consisting entirely of *hydrogen and *carbon. The simplest hydrocarbons are *methane, CH_4, and propane C_2H_6, which are both *alkanes. Other organic compounds may contain *oxygen, *nitrogen, or *sulphur and are hence not strictly hydrocarbons, but the term is loosely used in the oil industry to describe all of the components of *crude oil and *natural gas.

hydrogen The lightest element, which is a gas; symbol H. It occurs in small amounts in *natural gas accumulations, although there are records of small pure hydrogen accumulations. It exists in three isotopic forms, common hydrogen, ^1H; deuterium, ^2H; and radioactive tritium, ^3H. It forms *hydrocarbons when combined with carbon. Hydrogen-rich *kerogens are *oil prone. *See also*: hydrogen isotopes.

hydrogen index A parameter derived from *pyrolysis which measures the hydrogen richness, usually of *kerogen. It is calculated from S_2 and *total organic carbon content values (S_2/TOC). It has a direct relationship with elemental *H/C ratio (see Fig., p. 71). The index is measured in units of mg hydrocarbon/g organic carbon. For immature kerogen a hydrogen index of 600 to 950 indicates *Type I kerogen; 400 to 600, *Type II kerogen; and 0 to 300 *Types III and IV (or IIIB) kerogen. At high levels of maturity, kerogen types become indistinguishable because their chemical compositions are similar. If the *kerogen type is known, then the hydrogen index can be used to indicate the maturity of kerogen. *See also*: oxygen index, H/C ratio.

hydrogen isotopes Hydrogen (H) and deuterium (D) isotopic abundances are used in the analysis of *hydrocarbon gases, especially *methane, to define their maturity and origin (see Fig. to illustrate ranges of values for geological materials). The abundance of deuterium is measured relative to a standard, *SMOW (Standard Mean Ocean Water). Deuterium enrichment occurs on increasing *maturation. In water, hydrogen isotopic *fractionation reflects water temperature. Radioactive tritium, with a half life of 12.5 y, is used for isotopic

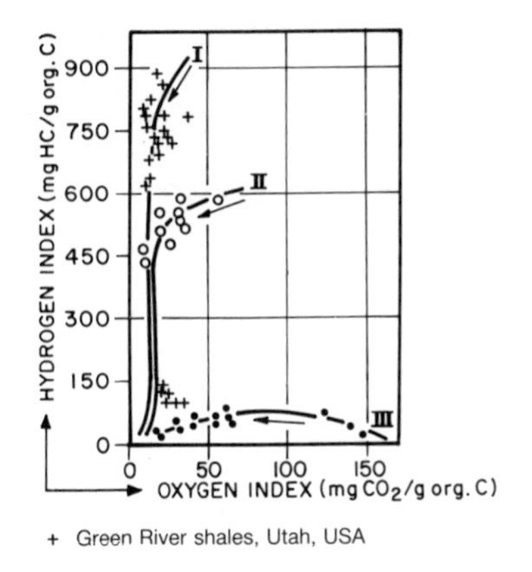

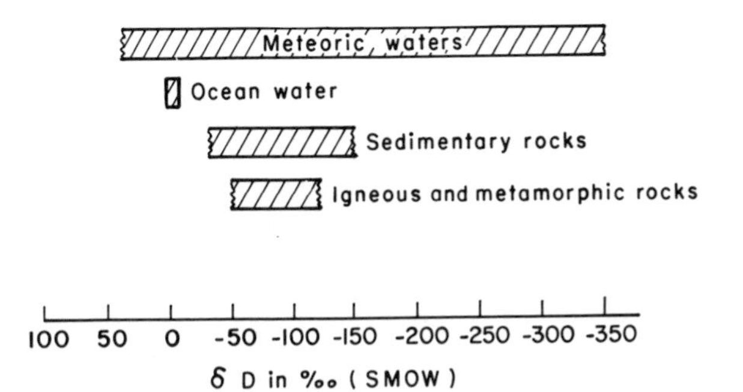

labelling in short-lived tracer experiments. *Reference*: Estep, M.R. and Hoering, T.C. (1980). Biogeochemistry of the stable hydrogen isotopes. *Geochim. et Cosmochim. Acta*, **44**, 1197–206.

hydrogen sulphide A poisonous gas smelling of rotten eggs, formula H_2S. It occurs naturally as a decomposition product of organic material in the presence of *sulphur and can be a constituent of natural oil and gas accumulations. It is particularly insidious because it cannot be detected after even short exposure, as it destroys the sense of smell. If found in oil and gas accumulations it is called sour gas. It is a drilling hazard and logging units are equipped with sophisticated detectors for monitoring its presence.

hydrous pyrolysis *Pyrolysis in the presence of water, usually in a sealed unit known as a bomb. Hydrous pyrolysis may be used to simulate natural *maturation processes and to measure *activation energies of *kerogen. It gives closer results than anhydrous pyrolysis to normal maturation products. It is mostly used to mature *source rocks artificially for *oil–source correlations. *Reference*: Lewan, M.D. (1985). Evaluation of petroleum generation by hydrous pyrolysis experimentation. *Phil. Trans. R. Soc. Lond.*, **A 315**, 123–34.

immature The *maturation stage where *kerogen has yet to generate significant amounts of oil or gas. It is broadly equivalent to *diagenesis. A source rock is immature with respect to oil generation at a *spore colour index of <3.5 (on a scale of 1 to 10), *thermal alteration index (Staplin) <2.2, and a *vitrinite reflectance of <0.5 per cent. *Gas prone sediments are said to be immature at *TAI <2.5 and vitrinite reflectance values of <0.7 per cent. The *early mature stage is the next higher maturation level. *See*, Table 1, summary section, for a summary of maturation parameters. *See also*: early mature, peak mature, late mature, post mature.

impsonite *See* solidified bitumen.

inclusion complex *See* adduction.

inertinite A *coal petrological term used in the microscopic study of *kerogen to describe the *maceral group which has no hydrocarbon potential (also known as *dead carbon). The group contains *fusinite, *semifusinite, *inertodetrinite, *macrinite, *micrinite, and *sclerotinite, which are particles which have high reflectance. They are carbon-rich, and oxidized either by low-temperature processes or by fire. They equate chemically to *Type IV kerogen. *See also*: fusinite, sclerotinite, semifusinite.

inertodetrinite Small fragments of *fusinite or *semifusinite.

infra-red spectroscopy (IR) Organic compounds absorb light in the infra-red region of the spectrum. The absorption can generally be correlated with the presence of various functional groups –OH, =C=O, –C=C–, substituted and unsubstituted aromatic rings. This analysis is called infra-red spectroscopy. *References*: Organic chemistry textbooks.

ingramite *See* solidified bitumen.

inverse modelling A technique for calculating the thermal history of a basin, or horizon within a basin, through time using chemical kinetics, the inverse of *maturation modelling. Successful application of the technique requires that the *order of the reaction, *activation energy, *A factor, and the progress (percentage conversion) of sets of reactions are known. The method has mainly been applied in *organic geochemistry using *amino acids, *aromatic steroids, *triterpanes, and *vitrinite reflectance changes, where the *kinetics of the various reactions are thought to be reasonably well understood. The method has also been applied using inorganic reactions such as apatite fission track annealing. *References*: Lerche, I., Kendall, C., and Yarzab, R.F. (1984). The determination of palaeoheatflux from vitrinite reflectance data. *Am. Assoc. Petr. Geol. Bull.*, **68**, 1704–17. Mackenzie, A.S. and McKenzie, D.P. (1983). Isomerization and aromatization of hydrocarbons in sedimentary basins formed by extension. *Geol. Mag.*, **120**, 417–70.

irregular isoprenoids Isoprenoids formed by 'head to head' or 'tail to tail' linked *isoprene units. Examples are the *carotenoids, lycopane, and squalane.

isoalkanes Specifically, isoalkanes are 2-methylalkanes (straight-chain alkanes with a methyl group on the second member of the carbon chain). Isoalkanes are derived from plants and bacterial waxes. C_{23} and C_{27} isoalkanes are believed to be sourced by halophilic *bacteria. Isoalkane to *normal alkane indices are used as *maturation parameters. *See also*: butane, isoheptane value.

isoheptane value Isoheptane is the *branched alkane containing seven carbon atoms. It has several *stereoisomers and their relative abundance is used to calculate the source and maturity index derived from *gasoline range hydrocarbon analysis of oils, *condensates and *source rocks, known as the isoheptane value. It is calculated from the formula:

$$\frac{\text{2-methylhexane} + \text{3-methylhexane}}{cis\text{-1,3-DMCP} + trans\text{-1,3-DMCP} + trans\text{-1,2-DMCP}}$$

where DMCP is dimethylcyclopentane. A value of 0.8 to 1.2 represents normal maturity, 1.2 to 2.0 mature, 2.0 to 4.0 super-mature, 0.0 to 0.8 biodegraded. The illustration shows how the isoheptane and heptane

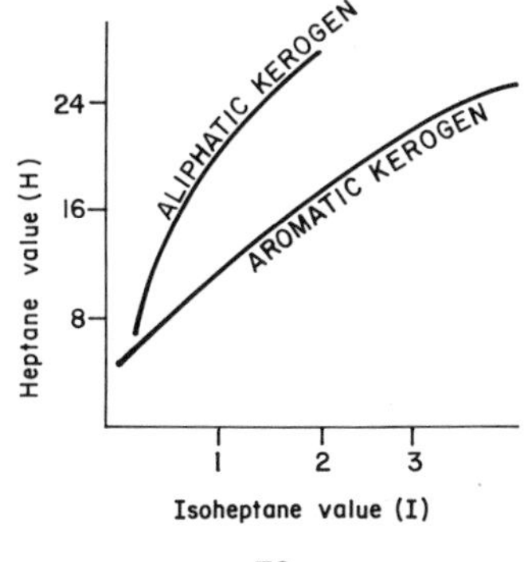

values can be plotted to determine *kerogen type of the source organic matter. *See also*: heptane value. *Reference*: Thompson, K.F.M. (1983). Classification and thermal history of petroleum based on light hydrocarbons. *Geochim. et Cosmochim. Acta*, **47**, 303–16.

isomer When a number of molecules fit a given molecular formula they are known as isomers. There are several different types of isomerism: *structural, positional, *optical, and *stereoisomerism. An example of structural isomerism is normal- and iso-butane (see Fig.). Positional isomerism occurs when there are several different sites for a substituent group, as in normal propanol and isopropanol (see Fig.). Optical isomers occur when four different substituents on, e.g. a carbon atom, produce two, non-superimposable, mirror images of the molecular arrangement (*R and *S). This form of isomerism is important in *organic geochemistry as the dominance of one of these isomers can only be produced biologically. Optical and stereoisomers may change from one form to another as a result of thermodynamic factors. *See also*: optical isomers, stereoisomers, structural isomers. *References*: Organic chemistry textbooks.

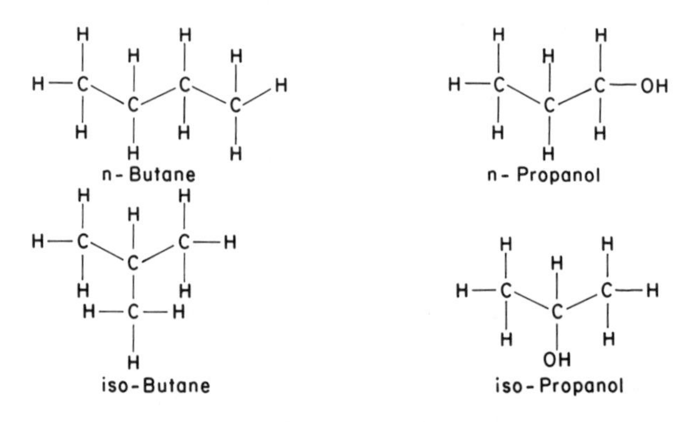

isoprene A prominent biochemical building block for terpenoids consisting of five carbon atoms which may become linked by conjugated double bonds. The isoprene unit is described as having a 'head' and a 'tail' (see Fig.). The normal linkage for multiples of the unit in rings or chains, is called 'head to tail' and gives rise to *regular isoprenoids such

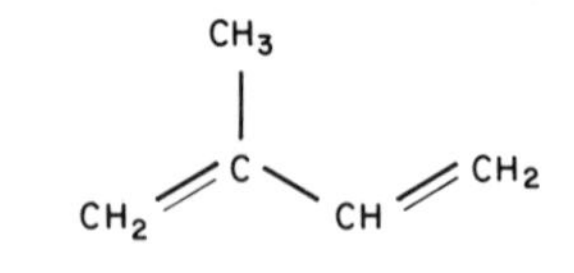

Isoprene unit

as *pristane. 'Head to head' and 'tail to tail' linkages give rise to *irregular isoprenoids, such as squalane. The monomer, isoprene is rare in nature, but multiples are common. They are classified as follows: two units, (mono) terpenes (C_{10}) in higher plants and *algae; three units, sesquiterpenes (C_{15}) in plants; four units, diterpenes (C_{20}) in higher plants; six units, triterpenes and *steroids (C_{30}) in *bacteria, algae, and higher plants; eight units, tetraterpenes (C_{40}) in *carotenoid pigments in higher plants.

isoprenoids Isoprenoids are compounds such as *hydrocarbons, alcohols, and esters built from multiples of the *isoprene unit. They are dominantly derived from plant and bacterial sources, and are *biological markers. The compounds may be either cyclic or acyclic in structure. Two of the most well known *acyclic isoprenoids are *pristane and *phytane, which are products of the side chain of the *chlorophyll molecule and are members of the series known as *regular isoprenoids. C_{14} to C_{20} compounds are often detectable by *gas chromatography, but higher numbers may be detected by *gas chromatography–mass spectrometry, with a *diagnostic ion of *m/e* 183.

Different isoprenoids are environmental indicators, C_{25} was believed to be derived from saline lagoonal sources, and high molecular weight isoprenoids from bacterial sources. *Botryococcanes, a series of compounds from C_{31} to C_{33}, are indicative of a *non-marine algal source. The isoprenoids are used in *oil to oil and *oil–source correlations. *See also*: irregular isoprenoids, phytane, pristane, regular isoprenoids.

isotopes Isotopes are different nuclear forms of the same element, with the same atomic number but slightly different atomic weights. In some elements this produces an unstable nucleus with radioactive decay. Each isotope has a characteristic decay time known as the half life. These isotopes may be used in age dating, e.g. ^{14}C. Stable isotopes are not subject to this decay and their abundance is controlled by other processes, e.g. ^{13}C by biological processes, ^{18}O in water by temperature. C, H, O, and S have stable isotopes, whose relative abundances are used for various purposes in *organic geochemistry. The relative abundance of ^{12}C to ^{13}C may be a useful parameter in determining *source rock depositional environments and for *oil to oil and *oil–source correlations.

Karweil diagram A diagram relating the relative effects of temperature and time to the *maturation of coal (see Fig., p. 76). The axes were temperature in °C and Z scale. Formation ages were in my. Maturity

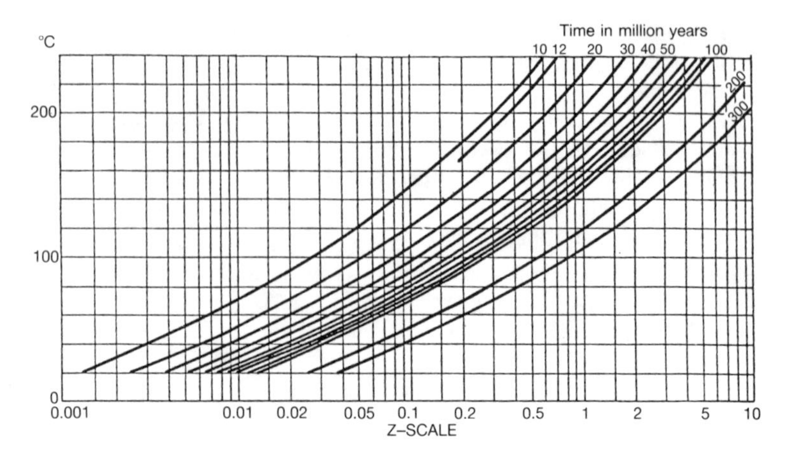

was expressed as per cent volatile matter. It was one of the original attempts to quantify the contribution of time in the *maturation process. This was published in 1956 by J. Karweil in German in *Zeitschrift der Geologischen Gesellschaft*, **107**, 132–9.

katagenesis *See* catagenesis.

kerogen Insoluble *organic matter which is preserved in sedimentary rocks. It derives from the breakdown and *diagenesis of plant and animal matter. Kerogen may be classified visually into groups known as *macerals, on the basis of morphology, colour in white and UV light, and associated particles. It may be classified chemically on the basis of its C, H, and O elemental composition. Kerogen may be isolated from sediments by hydrofluoric and hydrochloric acid attack to remove the mineral matrix, or by density separation using heavy liquids. Under the increasing influence of temperature and time (*maturation), most kerogen produces *hydrocarbons. The precursors of the kerogen, and degree of preservation determine the chemistry of the kerogen and hence the amount and type of hydrocarbon products. Kerogen is destroyed by oxidation. *See also*: kerogen type. *Reference*: Durand, B. (ed.) (1980). *Kerogen—insoluble organic matter from sedimentary rocks*, 519 pp. Editions Technip, Paris.

kerogen type The classification of *kerogen on the basis of visual properties or chemical composition is known as its kerogen type. The major kerogen types on the basis of optical properties are: *liptinite or *exinite, *vitrinite, and *inertinite (described by microscopic organic analysis). On the basis of C, H, and O elemental composition four *Types are recognized, *I, II, III, and IV (IIIB). Exinites or liptinite, Type I and Type II kerogen, produce oil when *mature, and gas when *post mature. Vitrinite and Type III kerogen produce gas when mature.

Inertinite and Type IV (IIIB) kerogen do not produce significant hydrocarbons at any level of maturity.

kinetics *See* chemical kinetics.

lacustrine Pertaining to, produced by, or formed in a lake. Lacustrine source rocks are those which contain *kerogen derived from non-marine algal, bacterial, and terrigenous organic matter. They frequently contain *alginite or *Type I kerogen, e.g. Green River Formation, USA. *Extracts of the sediments and oils derived from them often have a uni- or bimodal *alkane distribution, indicating algal and land plant source. These typically peak at C_{15} to C_{19} for the algal component, and C_{25} to C_{31} for the land plant component. They are characterized by low *sterane to *hopane ratios, and prominent C_{28} and C_{29} steranes. Sediments from saline lakes often contain high relative abundances of *methyl steranes and *gammacerane.

Salinities in lacustrine settings may vary from fresh through brackish, saline, and mesosaline to hypersaline. Lakes may also be acid or alkaline, and reducing or oxidizing. The differences in these physical conditions lead to differences in the preservation of organic matter and in trace metal concentrations, which allow these conditions to be defined. *Nickel/vanadium ratios may be used to separate oils derived from organic matter deposited in acid or alkaline lakes. The *stable carbon isotope ratios of lacustrine organic matter may be isotopically lighter than *marine organic matter. Synonymous with limnic. *See also*: carotenoids, gammacerane, non-marine. *Reference*: Powell, T.S. (1986). Petroleum geochemistry and depositional setting of lacustrine source rocks. *Mar. Petr. Geol.*, **3**, 200–19.

late mature The stage of maturity when oil or gas generation is waning, between *peak and *post mature. For *oil prone *source rocks this equates to a *spore colour index of 7 to 9 (scale 1 to 10), *thermal alteration index (Staplin) 2.6 to 3.5, and *vitrinite reflectance 1.0 to 1.3 per cent. For *gas prone source rocks, this equates to a TAI of 3.6 to 4.0 and vitrinite reflectance of 2.2 to 3.0 per cent. *See*, Table 1, summary section, for maturation summary. *See also*: early mature, immature, peak mature, post mature.

Level of Maturity *See* LOM.

light hydrocarbons Those *hydrocarbons which are gaseous and near gaseous at normal temperatures and pressures are called light hydrocarbons. They range from *methane (C_1) to octane (C_8), including *aromatic compounds, *normal, *iso-, and *cyclic alkanes. Their abundance and composition are used to detect the presence of *source rocks and reservoired hydrocarbons in *cuttings and *headspace gas analysis. *See also*: butane, heptane value, isoheptane value. *Reference*: Thompson, K. F.M. (1979). Light hydrocarbons in subsurface sediments. *Geochim. et Cosmochim. Acta*, **43**, 657–72.

light (isotopes) Containing relatively more of the lighter of two stable *isotopes, e.g. *biogenic methane is depleted in ^{13}C relative to ^{12}C so is said to be isotopically lighter than *thermogenic methane. *See also*: biogenic gas.

light oil An oil with an *API gravity of between 35 and 45° which has a white *fluorescence in UV light. It is the product of *mature to *late mature *exinite or *liptinite. Mixed oil and gas prone assemblages may also produce light oils with a chemistry similar to that of *condensates.

lignin The natural *carbohydrate polymer which is a constituent of woody material, and gives it physical strength. It is a precursor to *vitrinite or *huminite.

lignite A brownish black, low rank *coal, between peat and sub-bituminous coal, and of reflectance 0.2 to 0.4 per cent. It is commonly used as a drilling mud additive. It causes contamination of geochemical samples, and may lead to erroneous determinations of *total organic carbon, *vitrinite *reflectance, and *pyrolysis results unless identified and removed.

limnic *See* lacustrine.

lipids A diverse group of compounds including *waxes, *alkanes, *fatty acids, and esters, defined by their solubility in chloroform, *benzene, ethers, and acetone. They are classified as simple lipids, e.g. fats; compound lipids, e.g. cell membranes; and non-saponifiable lipids, e.g. cholesterol. Most lipids are composed entirely of C, H, and O, but some may contain N and P. They have neither a common structure nor functional group, although the majority are esters derived from long chain carboxylic acids (fatty acids). They are found in the exines of plant spores, seeds, and fruit (where they are an energy store), as a protective coating on leaves, and in *algae and *bacteria. They are a constituent of *kerogen which produces oil, which is called *liptinite.

liptinite The *maceral group which is derived from organic matter which is high in lipids. It is distinguished microscopically by its structure, and its *fluorescence in UV light. It is divided into two sub-groups, unstructured and structured *liptinite. Structured liptinites are *sporinite, *suberinite, *alginite, and *cutinite. Unstructured liptinites are *fluorinite, *bituminite, *exsudatinite, *liptodetrinite, and *resinite. It is synonymous with *amorphous (algal) *kerogen, *herbaceous (structured) kerogen, *sapropel, and *exinite. It equates chemically with *Type I and *Type II kerogen.

liptodetrinite The collective term for unstructured *liptinite constituents of different form, low reflectance and *fluorescence, which are detrital and cannot be assigned to the liptinite *maceral groups. It may comprise fragments of spore, cuticle, resin, or *algae, and is characteristic of *coals deposited under water. Particles can be as small as clay size. It often occurs as grains, rodlets, fibres, and splinters embedded in *bituminite. It equates chemically with *Type II kerogen.

liquid chromatography The separation of liquid mixtures of compounds on the basis of a combination of their polarity, retention on different stationary phases, and solubilities in different solvents. In *organic geochemistry, *crude oils and sediment *extracts are subdivided by this technique into their *saturate, *aromatic, *polar, or *NSO fractions. Three versions of the method in use in organic geochemistry are *HPLC, *TLC, and *column chromatography.

LOM 1. A widely used abbreviation for the Level of Organic Maturity.
2. A maturity scale from 1 to 20 based on *vitrinite reflectance published by *Hood *et al.* An LOM of 7.8 equates to onset of oil generation, 9.5 to peak oil generation, and 11.5 to the end of oil generation. It is also used as a scale for *theoretical maturity calculations in the method of Hood *et al. Reference*: Hood, A., Gutjar, C.C.M., and Heacock, R.L. (1975). Organic metamorphism and the generation of petroleum. *Am. Assoc. Petr. Geol. Bull.*, **59**, 986–96.

Lopatin The author of a technique for calculating the theoretical maturity of *source rocks. The method was adapted by Waples (1980) for oil exploration. The method allows calculation of the interchangeable effects of temperature and time on *maturation of *kerogen. *Chemical kinetics predicts that the time dependence of maturation is linear whilst the temperature dependence is exponential. In order to be able to take these two dependencies into account simply, the temperture history is divided into 10°C steps. A *burial history is constructed and, using a temperature–depth profile, the time that an horizon or source rock spends in each 10°C segment, from 30°C to its maximum temperature, is calculated. This time is then multiplied by a factor of two; the factor depends on the absolute temperature increment (these increments and factors of two are listed below). This gives a *TTI (Time Temperature Index), which is summed up over the whole rock history. In the absence of any real maturity data to calibrate the modelled maturity, a value of TTI of 15 is equivalent to the onset of oil generation, 75 is peak oil generation, 160 the end of oil generation, and 1500 the wet gas deadline. The *activation energy that the method uses is ~20 kcal mol^{-1} at 100°C. The method appears to be reasonably accurate at low levels of maturity.

Indices are: 30 to 40°C, 2^{-7}; 40 to 50°C, 2^{-6}; 50 to 60°C, 2^{-5}; 60 to 70°C, 2^{-4}; 70 to 80°C, 2^{-3}; 80 to 90°C, 2^{-2}; 90 to 100°C, 2^{-1}; 100 to

110°C, 2°=1; 110 to 120°C, 2; 120 to 130°C, 4; 130 to 140°C, 8; 140 to 150°C, 16; 150 to 160°C, 32; etc. *See also*: activation energy, Hood *et al.*, pseudoactivation energy. *Reference*: Waples, D. (1980). Time and temperature in petroleum formation: application of Lopatin's method to petroleum exploration. *Am. Assoc. Petr. Geol. Bull.*, **64**, 916–26.

lupanes *See* triterpanes

M−1, M−15 Abbreviation for the *molecular ion with the loss of 1 *m/e* or 15 *m/e* units.

m/e, m/z The mass to charge ratio of fragments of molecules, especially from *gas chromatography–mass spectrometry. *See also*: diagnostic ion.

maceral Coalified plant remains possessing distinctive chemical and physical properties which change with *rank. They are also, in part, degradation products of plants whose origins can no longer be recognized. The term is analagous to a mineral in describing microscopically recognizable components of *kerogen. Three main divisions of macerals are made, into *liptinite, *vitrinite (or *huminite), and *inertinite, equating to *oil prone, *gas prone, and inert kerogen. *See also*: inertinite, liptinite, vitrinite. *Reference*: Murchison, D., Cook, A.C., and Raymond, A.C. (1985). Optical properties of organic matter in relation to thermal gradients and structural deformation. *Phil. Trans. R. Soc. Lond.*, **A 315**, 157–86.

macrinite *Humic tissue which has been gelified then oxidized, similar to *semifusinite. It is more or less amorphous, non-granular and occurs as a ground mass with other *macerals embedded in it. It equates chemically to *Type IIIB or *IV kerogen.

marine organic matter Marine organic matter is characterized by having abundant *normal alkanes, generally C_{12} to C_{20}, with algal-derived alkanes C_{15} and C_{17} frequently dominant. C_{15} to C_{20} *isoprenoids are common, and C_{27} *steroids are generally more abundant than C_{29} steroids. The abundance of C_{28} steroids has increased through geological time, probably as a result of increasing phytoplankton diversity. The relative abundance of C_{27}:C_{28}:C_{29} mono*aromatic steroids may also indicate whether the *hydrocarbons are derived from a marine or non-marine source. The occurrence of C_{30} steranes is thought to be a reliable marine indicator. *Triterpanes are present and *hopanes are commonly in equal abundance to *steranes. *Stable carbon isotope ratios are mostly between −22 and −30 per mil (*PDB).

mass spectrometry *See* gas chromatography–mass spectrometry.

maturation The process of chemical change in sedimentary organic matter, induced by burial, i.e. the action of increasing temperature and pressure over geological time, is known as maturation. These chemical changes produce oil and *hydrocarbon gases from the appropriate organic matter. Maturation is either measured directly by the changes in chemistry of the *kerogen or the soluble organic matter, or indirectly from resulting colour and reflectivity changes. The major maturity subdivisions are *immature, *early mature, *peak mature, *late mature, and *post mature. *See*, Table 1, summary section, for maturation summary. *See also*: early mature, immature, late mature, peak mature, post mature. *Reference*: Heroux, Y., Cagnon, A., and Bertrand, R. (1978). Compilation and correlation of major thermal maturation indicators. *Am. Assoc. Petr. Geol. Bull.*, **62**, 2128–44.

maturation modelling The use of basic laws of *chemical kinetics to estimate the level of maturity of source rocks in the absence of measured maturity data. This may be used for locations in accessible parts of a basin prior to drilling, or for extrapolation into inaccessible parts of a basin. The *maturation levels are calculated from thermal models, or measured temperature information and some form of the *Arrhenius rate equation. Depending upon the type of data available, the Arrhenius equation may be used in its accurate form with *activation energies, e.g. *Tissot and Espitalié Model, or using approximations to the Arrhenius equation which do not require activation energies, e.g. *Lopatin and *Hood *et al.* Ideally, results from maturation models should be calibrated against measured maturity data before use, but this may not be practical in undrilled areas. Most models have default values which are of broad applicability for use in these circumstances. Maturation models may be integrated with thermal, rock compaction, and fluid flow models to provide estimates of the timing of oil generation and *migration, in addition to simple estimates of maturity (see Fig., p. 82). *See also*: Hood *et al.*, Lopatin, Tissot and Espitalié Model. *Reference*: Waples, D.W. (1984). Thermal models for oil generation. In *Advances in Petroleum Geochemistry I* (ed. J. Brooks and D.H. Welte), 344 pp. Academic Press, London.

mature A mature sediment is one which has reached sufficient thermal alteration to generate *hydrocarbons. For an *oil prone source rock, this would be a *spore colour index of 3.5 to 9.0 (scale of 1 to 10), *thermal alteration index (Staplin) 2.2 to 3.5, *vitrinite reflectance of 0.5 to 1.3 per cent. For a *gas prone source rock 0.7 to 3.5 per cent vitrinite reflectance. Three levels of maturity are recognized, *early, *peak, and *late mature. *See*, Table 1, summary section for maturation summary. *See also*: early mature, late mature, peak mature.

medium gravity oil A medium gravity oil is one whose *API gravity is

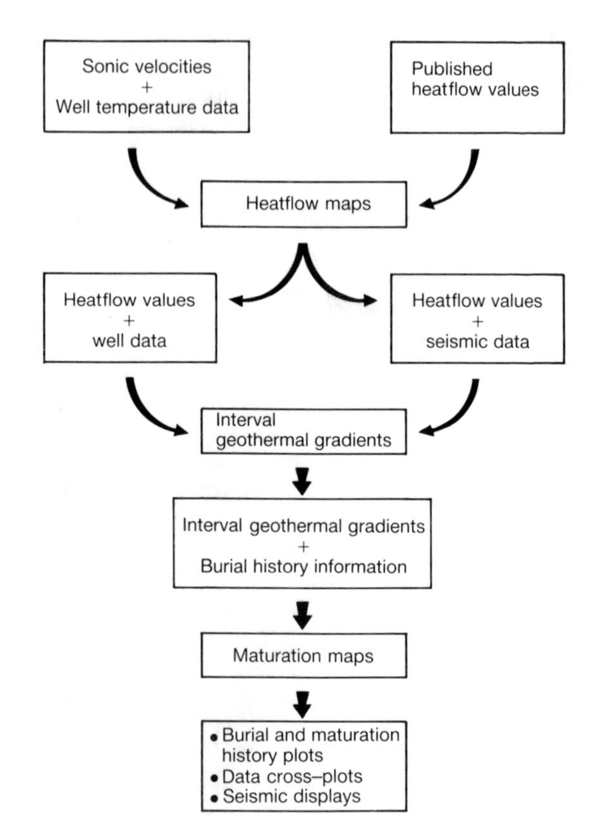

from 25° to 35°, and *fluorescence colour yellow or cream. It is generally the product of *oil prone source rocks at moderate to high levels of maturity.

metagenesis A high level of *maturation spanning the gap between inorganic and organic metamorphism, where *methane is the only *hydrocarbon product. *Kerogen becomes more aromatic in nature, forming sheets of ring clusters and approaching graphite structure. Mineral transformations begin, and eventually the zone of greenschist grade metamorphism is reached. It is equivalent to a range of between 2 and 4 per cent *vitrinite reflectance. *See also*: catagenesis, diagenesis.

metastable ions The name given to ions of intermediate stability formed in a region known as the first field-free region of a double focusing mass spectrometer (*GC–MS–MS), between the electron source and the magnet which separates ions on the basis of their mass to charge ratio. The presence of a metastable peak confirms the route by which a fragmentation occurs, from *parent ion to *daughter ion. They are used when, for example, isomers co-elute in conventional *gas chromatography–mass spectrometry.

Their most successful application has been in *sterane geochemistry, where isomers of *methyl, *regular, and *rearranged steranes all elute together, so that the calculation of the relative abundance of different isomeric forms for source and maturity determination is only possible for the C_{29} steranes. They are also used in the analysis of C_{30} steranes, which are believed to be indicators of *marine organic matter. The transition of, for example, ions of *m/e* 386 (C_{28} steranes) to the *m/e* 217 daughter fragment, can be examined on its own so that other types of steranes do not interfere. The *SIM and *MID techniques are replaced by Selective Metastable Ion Monitoring or *SMIM. *Reference*: Warburton, G.A. and Zumberge, J.E. (1983). Determination of petroleum sterane distribution by mass spectrometry with selective metastable ion monitoring. *Anal. Chem.*, **55**, 123–6.

methane The simplest lightest and most abundant *hydrocarbon, formula CH_4; it is a gas at room temperatures (see Fig.). It is produced from all (except heavily oxidized) organic matter. During *diagenesis methane is produced by *bacteria; during *catagenesis, it is produced in large quantitites from the thermal decomposition of *liptinite and *vitrinite; and during *metagenesis it is produced in minor amounts from all but the most oxidized *kerogen. It is the major constituent hydrocarbon in *natural gas accumulations, and can occur as free gas or dissolved in oil accumulations. It complexes with water under a narrow range of temperatures and pressures to form *gas hydrates. *Reference*: Schoell, M. (1980). The hydrogen and carbon isotopic composition of methane from natural gases of various origins. *Geochim. et Cosmochim. Acta*, **44**, 649–61.

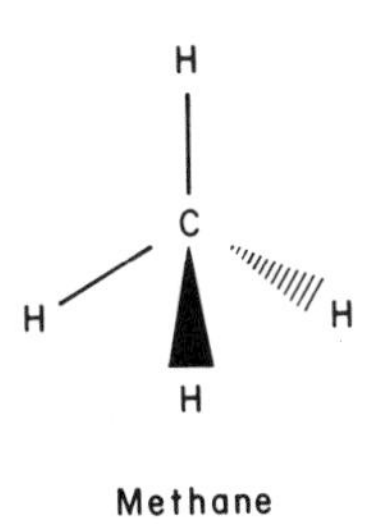

Methane

methanogenic bacteria *Anaerobic bacteria which produce *methane by fermentation of organic matter are known as methanogenic bacteria. They are active in surface and near-surface sediments and many other natural environments. The methane is produced by the reduction of acetate or carbon dioxide. Methanogenic bacteria are competitive with *sulphate-reducing bacteria, and tend not to prosper in sediments until sulphate has been depleted. *Fractionation of both *carbon and *hydrogen isotopes occurs, making the methane produced distinctive. *See also*: sulphate-reducing bacteria. *Reference*: Brassell, S.C., Wardroper, A.M.K., Thompson, I.D., Maxwell, J.R., and Eglinton, G. (1981). Specific acyclic isoprenoids as biological markers of methanogenic bacteria in marine sediments. *Nature*, **290**, 693–6.

methylbenzene *See* toluene.

methylcholestane C_{28} *sterane. Synonymous with ergostane.

Methylphenanthrene Index (MPI) In sediment *extracts from rocks containing hydrogen-poor *kerogen, the concentrations of 2- and 3-methylphenanthrenes (shorthand for phenanthrene is 'phen' and for methylphenanthrene is 'me') are seen to increase as a result of increasing *maturation. It is believed that this occurs as a result of rearrangement of 1- and 9-methylphenanthrenes, as the 2- and 3- forms are thermodynamically favoured at higher temperatures. Two indices have been proposed, known as MPI 1 and MPI 2. These are calculated as follows:

$$\text{MPI 1} = \frac{1.5\ (\text{2-me} + \text{3-me})}{\text{phen} + \text{1-me} + \text{9-me}} \text{ and MPI 2} = \frac{3\ (\text{2-me})}{\text{phen} + \text{1-me} + \text{9-me}}.$$

The MPI is dependent on both facies and *kerogen type. The MPIs both reach maxima at *vitrinite reflectances of 1.3 per cent, then decrease, so that there are two values of R_o for one of each MPI (see Fig.). An MPI of 0.45 (0.6 per cent R_o) to 1.5 (1.3 per cent R_o) defines the *oil window. These values have also been recalculated as methylphenanthrene distribution factor (or MPDF), which is 2+3 me/1+2+3+9 me, which gives a linear correlation with vitrinite reflectance. *Reference*: Radke, M. and Welte, D.M. (1983). The methylphenanthrene index: maturity parameter based on aromatic hydrocarbons. In *Advances in Organic Geochemistry 1981* (ed. M. Bjorøy *et al.*) John Wiley, Chichester.

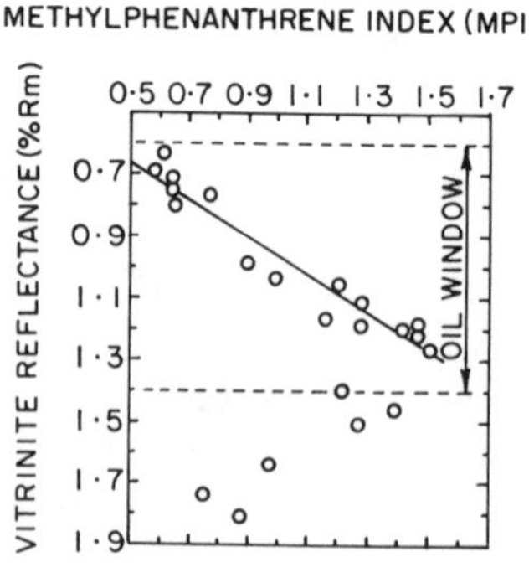

methyl steranes *Steranes with an additional methyl group, usually at the C-4 position (see Fig.). Their presence in oils and *source rock extracts is mostly attributable to a dinoflagellate source, but as they occur in both marine and non-marine rocks before the respective first appearances of dinoflagellates in marine (Permian) and non-marine (Late Cretaceous) settings, there must be an alternative, probably bacterial, source. High relative abundance of these compounds appears to occur with high abundance of the *triterpane gammacerane in saline lake sediments. They are present in the *saturate fraction of oils and *extracts and are analysed by *gas chromatography–mass spectrometry with a *diagnostic ion *m/e* 231. *Reference*: Boon, J.J., Rijpstra, W.C., deLange, F., de Leeuw, J.W., Yoshioka, M., and Shimizu, V. (1979). Black Sea sterols a molecular fossil for dinoflagellate blooms. *Nature*, **277,** 125–7.

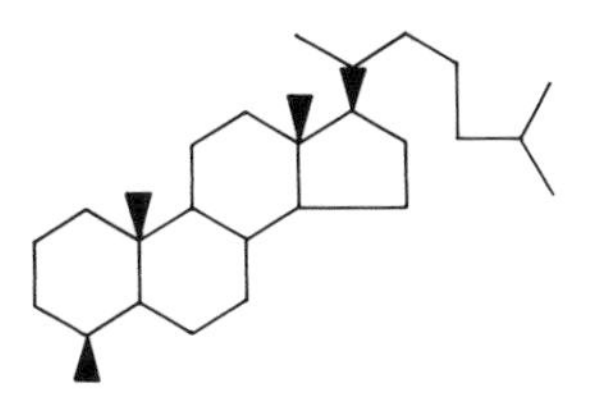

4-Methylcholestane

micrinite Granular *inertinite possibly formed as the residue from *kerogen after the generation of *hydrocarbons. It has a small grain size and particles tend to be rounded. It can occur as cell fillings in *vitrinite, and occurs commonly in *boghead coals.

MID Multiple Ion Detection. A mode of data collection during *gas chromatography–mass spectrometry, where only abundances of specified *m/e* fragments are monitored, instead of all ions.

migration Migration is the process whereby *hydrocarbons move from *source rocks to traps. Migration is divided into three categories: *primary migration where oil and gas leave the source rock, *secondary migration where oil and gas move in a *conduit from the source rock to the reservoir in trap configuration or are lost to the environment, and *tertiary migration where oil and gas move from one trap to another, or are lost (see Fig., p. 86). *See also*: primary migration, secondary migration, tertiary migration. *Reference*: Magara, K. (1980). Problems of petroleum migration. *Developments in petroleum science*, Vol. 9. Elsevier, Amsterdam.

migration pathway The route by which oil or gas move through *conduits from source rock to trap, i.e. by *secondary and *tertiary migration routes.

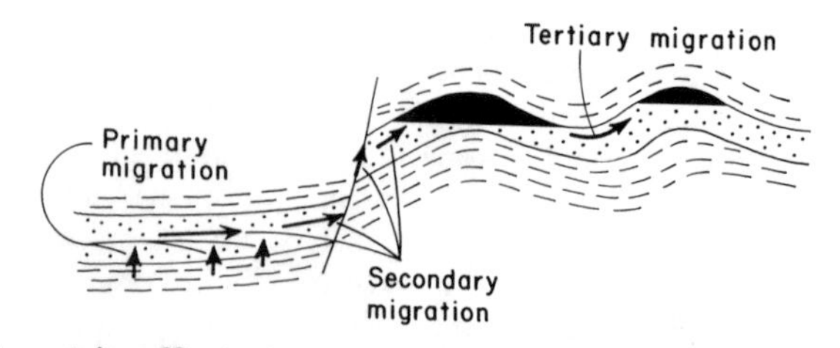

mineral matrix effect In some marginal *source rocks where *total organic carbon (TOC) contents are low, the matrix of the *source rock may retain a high proportion of the heavy *hydrocarbons of the S_2 yield during *pyrolysis. This results in a reduced S_2 yield and hence low *hydrogen index. Experiments have shown that this is caused by retention of pyrolysis products on the clays and carbonate of the rock matrix. Different clays have different affinities for hydrocarbons, and react in the order illite>montmorillonite>kaolinite. Calcite has also been shown to absorb pyrolysate. Up to half of the S_2 peak may be retained by the mineral matrix. Mineral matrix effects can be detected by comparing the hydrogen indices of isolated *kerogen with whole rock. *Reference*: Katz, B.J. (1983). Limitations of Rock Eval pyrolysis for typing organic matter. *Org. Geochem.*, **4**, 195–9.

MOA Microscopic Organic Analysis. *See* kerogen type.

molecular ions During *gas chromatography–mass spectrometry analysis, not all of the molecules of a compound fragment after ionization in the spectrometer. These unfragmented ions are detected at the *m/e* equivalent to the molecular weight of the compound, although the molecule has lost an electron, M^+. This is known as the molecular ion and it is important in compound recognition by GC-MS. Examples of some molecular ions are:

370 C_{27} triterpanes
384 C_{28} triterpanes
398 C_{29} triterpanes
412 C_{30} triterpanes
426 C_{31} triterpanes
440 C_{32} triterpanes
412 gammacerane

372 C_{27} steranes
386 C_{28} steranes
400 C_{29} steranes
414 C_{30} steranes
384 C_{28} methyl steranes
398 C_{29} methyl steranes
412 C_{30} methyl steranes

molecular sieve A substance such as a *zeolite which allows separation of compounds based on molecular size. Separation is achieved as smaller molecules become trapped in the crystal lattice of the molecular sieve. The lattice size can be varied for different classes of compounds. Commercial products with sizes from 5 to 13 Å are readily available. *See also*: adduction, zeolite.

moretanes C_{27} to C_{35} hopane-like compounds of the pentacyclic *triterpane series with 17β(*H*), 21α(*H*) stereochemistry (see Fig., p.

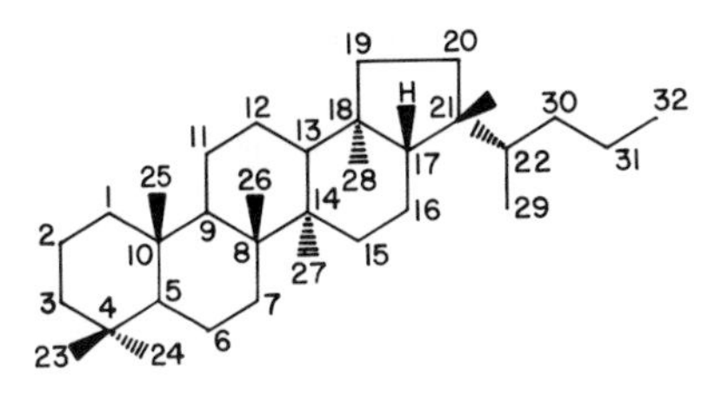

Hopane / Moretane

87). They are found only in sediments and oils, and their abundance decreases as a result of increasing *maturation. They are more abundant in coaly-derived organic matter. They are derived from 17β(*H*), 21β(*H*)-*hopane isomers. They are identified by several different *m*/*e* *diagnostic ions, from 177 upwards by increments of 14, during *gas chromatography–mass spectrometry analysis of the *saturate fraction of oils and *extracts.

MPDF *See* Methylphenanthrene Index.

MPI *See* Methylphenanthrene Index.

Multiple Ion Detection *See* MID.

n-alkanes *See* normal alkanes.

naphthalene A simple aromatic compound containing two *benzene rings (see Fig.). It is a liquid at normal temperatures, and is a constituent of most *crude oils.

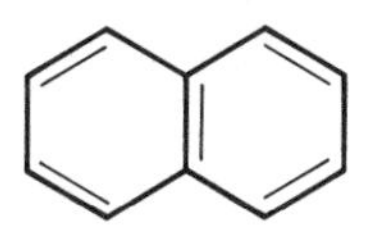

Naphthalene

naphthenes *See* cyclic alkanes.

natural gas A fossil fuel comprising on average, 60 to 80 per cent *methane, 5 to 19 per cent *ethane, 3 to 18 per cent *propane, 2 to 14 per cent C_4+, and various amounts of *nitrogen, carbon dioxide, *hydrogen, *hydrogen sulphide, and helium. The energy content of natural gas is normally 900 to 1300 BTU/SCF (British Thermal Units/Standard Cubic Foot). It is generally odourless; a sulphur compound is added so that it can be detected easily. *Reference*: Schoell, M. (1983).

Genetic characterisation of natural gases. *Am. Assoc. Petr. Geol. Bull.*, **67**, 2225–38.

NBS-18, NBS-19 Alternative standards to *SMOW and *PDB for *oxygen isotope analysis.

NBS-22 The US National Bureau of Standards petroleum standard used as an alternative to *PDB for *carbon isotopes. The two standards are related by the formula:

$$\delta^{13}C(PDB) = 0.9702 \times \delta^{13}C(NBS\text{-}22) - 29.8.$$

Reference: Schoell, M., Faber, E., and Coleman, M.L. (1983). Carbon and hydrogen isotopic compositions of NBS-22 and NBS-21. Stable carbon isotope reference materials, an interlaboratory comparison. *Org. Geochem.*, **5**, 3–6.

nickel/vanadium ratio, (Ni/V) The ratio of nickel to vanadium determined during refinery analysis of oils. The ratio depends on Eh/pH conditions in the water during organic matter growth and preservation, and hence reflects depositional environment. High Ni/V ratios are characteristic of alkaline *lacustrine environments (> 10:1), medium ratios, of acid lacustrine environments (10:1 to 1:1). Lower ratios are typical of marine environments. *Reference*: Lewan, M.D. (1984). Factors controlling the proportionality of vanadium to nickel in crude oils. *Geochim. et Cosmochim. Acta*, **48**, 2231–8.

nitrogen The gaseous element, N, atomic weight 14. It has two naturally occurring isotopes, ^{14}N and ^{15}N. Nitrogen gas is found occasionally as a major component of *natural gas accumulations. It is derived from the breakdown of *kerogen at high temperatures, as a product of *biodegradation, or from occluded atmospheric nitrogen in, for example, evaporites. It occurs in minor amounts in oils as compounds such as pyridine, quinoline, indols, carbonyls, and *porphyrins, and in sediments as *amino acids. Nitrogen compounds elute in the *NSO or *polar fraction of *crude oils.

non-marine Of lacustrine or terrestrial origin. Organic matter from these environments may be characterized by abundant C_{29} *steroids relative to C_{27} and C_{28}, low relative abundance of *sterane to *hopanes, the presence of higher land plant indicators such as *diterpenoids, sesquiterpenoids, *18α(*H*)-oleanane, bicadinane, compounds *X and *Y(triterpanes); and lacustrine indicators such as *botryococcane and *β-carotane. Stable carbon isotope ratios may be diagnostic if very light (−30 per mil or lighter). The absence of C_{30} steranes may also indicate a non-marine origin. *Reference*: Brassell, S.C., Eglinton, G., and Fu Jia Mo (1986). Biological marker compounds as indicators of depositional history of the Maoming oil shale. *Org. Geochem.*, **10**, 927–41.

25-norhopanes *See* demethylated hopanes.

normal alkanes *Straight-chain alkanes, abbreviation n-alkanes, of general formula C_nH_{2n+2}. They are found in organic matter, oils and sediments, where they may either be directly inherited from biological material or form from esters, alcohols, and *fatty acids during early *diagenesis. They are mainly derived from *algae, *bacteria, and land plants. Normal alkanes may form up to 30 per cent of *crude oils, and are eluted in the *saturate fraction during *liquid chromatography. They are usually the dominant compounds in the saturate fraction which can be demonstrated by *gas chromatography. Alkanes from C_1 to C_{40} are found in most oils and *extracts unless they have been removed by *biodegradation. Alkanes from some sources may show a natural predominance of odd-numbered over even-numbered forms. The reduction of this predominance occurs as a result of *maturation and can be used as an oil and source rock maturation parameter. Synonymous with paraffin.

A brief summary of the sources of alkanes in sediments and oils is:

Bacteria	C_{14}–C_{29}	dominant C_{17}, C_{25} and C_{26}	No *CPI
Fungae	C_{25}–C_{29}	dominant C_{29}	No CPI
Algae	C_{13}–C_{26}	dominant C_{17}, C_{23}– C_{29}	CPI
Land plants	C_{15}–C_{37}	dominant C_{27}, C_{29}, C_{31}	CPI

*Carbon Preference Index

See also: Carbon Preference Index, odd–even predominance. *Reference*: Brassell, S.C., Eglinton, G., Maxwell, J.R., and Philp, R.P. (1978). Natural background of alkanes in the aquatic environment. In *Aquatic Pollutants* (ed. O. Hutzinger, I. H. Van Lelyreld, and B. C. J. Zoeteman). Pergamon Press, Oxford.

N+P An abbreviation for naphthene and paraffin fraction of *crude oils. Synonymous with *saturate fraction and *alkanes.

NSO An abbreviation for *nitrogen, *sulphur, *oxygen, compounds of *crude oils. It is the fraction of crude oils which elutes with polar solvents, e.g. methanol in *liquid chromatography. Synonymous with *polar fraction, *hetero compounds, and 'heavy' fraction of oils. *See also*: column chromatography.

O/C ratio The atomic ratio of *oxygen to *carbon determined by *elemental analysis of organic matter or *kerogen. The O/C ratio may be plotted against *H/C ratio on a *Van Krevelen diagram to show *kerogen type and maturity differences. O/C elemental ratios are directly related to *oxygen indices from *Rock Eval *pyrolysis (see Fig., p. 90). *See also*: H/C ratio, Van Krevelen diagram.

odd–even predominance A *maturation parameter based on the

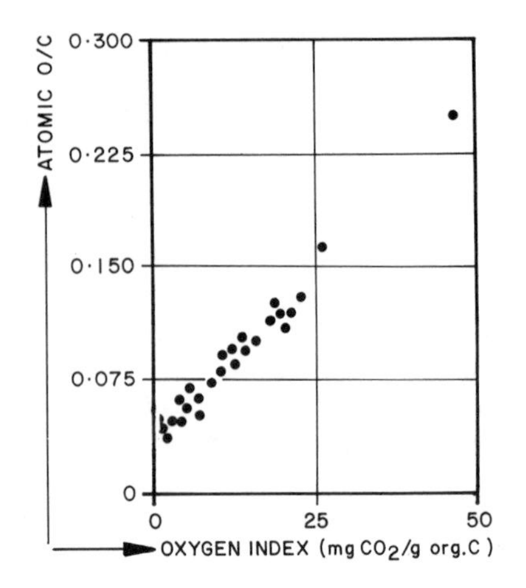

predominance of odd-numbered *normal alkanes in *immature sediments and oils. It was developed as an alternative to the *Carbon Preference Index to compensate for the different abundance of each alkane in the distribution. It is a running ratio over five carbon numbers as opposed to a fixed carbon number ratio, with the formula:

$$\left(\frac{C_i + 6C_{i+2} + C_{i+4}}{4C_{i+1} + 4C_{i+3}}\right)^{(-1)i^{-1}}.$$

See also: Carbon Preference Index. *Reference*: Scanlan, R.S. and Smith, J.E. (1970). An improved measure of the odd–even predominance in the normal alkanes of sediment extracts and petroleum. *Geochim. et Cosmochim. Acta*, **34**, 611–20.

oil–oil correlation The technique whereby oils are grouped on the basis of a common source, although the source itself is not necessarily known. Analyses which are used are *stable carbon isotope ratios, *gasoline range and *light hydrocarbon ratios, *gas chromatography, and *gas chromatography–mass spectrometry.

oil prone Organic matter which will generate oil at appropriate *maturation levels is described as oil prone.

oil shale An *immature, organic-rich, *oil prone *source rock, which can be retorted at ~500°C to give oil is called an oil shale. Oil shales typically contain 5 to 20 per cent organic carbon by weight and yield about 10 to 30 gallons of oil per ton, or 50 l per tonne of rock. Oil shales are an immense economic resource, but shale retorting is expensive. *In situ* retorting in underground locations is feasible, and has proved more economic than mining and retorting.

Oil Show Analyzer (OSA) Integrated *total organic carbon and *pyrolysis equipment which is used to detect migrated *hydrocarbons in reservoir rocks. It has *S_0 and *S_4 peaks (in addition to the *S_1 and *S_2 of *Rock Eval), which are equivalent to the released gases, and residual carbon or *TOC peaks, respectively. No *S_3 peak is recorded. The *production index of Rock Eval is subdivided into *Gas Production Index (GPI) and *Oil Production Index (OPI). *T_{max} is recorded as in Rock Eval.

oil–source correlation The correlation of an oil to its parent rock formation or *source rock is called an oil–source correlation. The major analyses which enable an oil to be correlated to its source are *stable carbon isotope ratios and *biological marker distributions. *Reference*: Schou, L., Eggen, S., and Schoell, M. (1985). Oil–oil and oil–source rock correlation, Northern North Sea. In *Petroleum geochemistry in exploration of the Norwegian Continental Shelf*, pp. 101–20. Norwegian Petroleum Society. Graham and Trotman, London.

oil window The maturity zone over which oils are generated from *source rocks is called the oil window. It equates to a *spore colour index range of 3.5 to 9 (scale 1 to 10) and a *vitrinite reflectance of 0.5 to 1.3 per cent, and it is *early to *late mature. *See* Table 1, summary section.

18α(*H*)-oleanane A *triterpane which elutes immediately before C_{30} *hopane during *gas chromatography–mass spectrometry analysis of the *saturate fraction of oils and *extracts. It has been circumstantially linked with a flowering plant (angiosperm) source. It has only been found so far, in Tertiary and Late Cretaceous-sourced oils. When found in oils it indicates a component of higher land plants contribution in the source organic matter.

OPI Oil Production Index. A parameter derived from the S_1 peak of the *Oil Show Analyser.

optical activity The property of certain compounds with asymmetric or *chiral centres, enabling them to rotate the plane of plane-polarized light either to the left or right. Some substances show optical activity only as a solid, (for example quartz), and some substances retain this property in the liquid and gaseous states. Optical activity can be exhibited by individual molecules rather than crystals. The magnitude and sign, − or +, and direction of rotation under specific conditions is known as the specific rotation. The rotation occurs when there is a dominance of one *optical isomer in the solution. Optical activity can only be the result of biological selection, as optical isomers have identical chemical properties, so when chemically produced are present in equal amounts. Compounds such as *amino acids and *steroids show optical activity. *References*: Organic chemistry textbooks.

optical isomers When an organic compound contains a carbon atom with four different substituent groups, there is one isomer which is the non-superimposable, mirror image of another (the figure shows two imaginary molecules to illustrate this), although they have identical chemical and physical properties. These two arrangements are known as optical isomers or *enantiomers, and they rotate the plane of plane-polarized light in opposite directions. If this rotation is to the left, the activity and isomers are known as – (negative) or laevo (L) rotatory, and to the right, + (positive) or dextro (D) rotatory. When the two

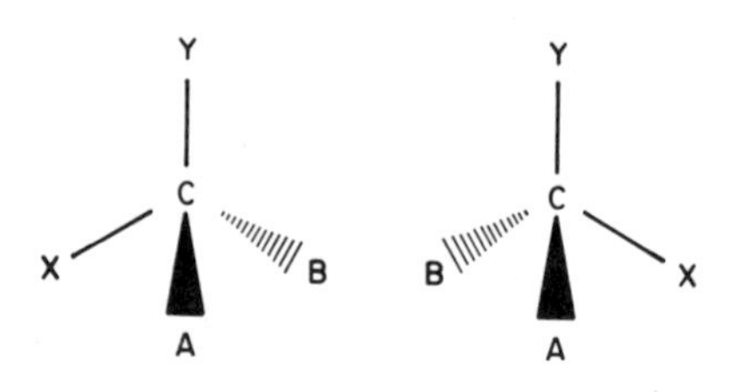

enantiomers are present in equal amounts they form a *racemic mixture. Simple optical isomers are said to have dextro and laevo rotatory forms, but when the molecules become more complex the two forms are called *rectus (R) and *sinister (S), according to a hierarchy of substituent groups. For many compounds there is a relationship between configuration and biological activity; for example, only D-glucose (not L) is fermented by enzymes found in most micro-organisms, and only L forms of *amino acids are incorporated into *proteins. *See also*: *R*, *S*.

order of a chemical reaction The general expression for the rate of a chemical reaction is in the form:

$$\mathrm{d}c/\mathrm{d}t = -k(c),$$

where c is the concentration, t is the time, and k the rate constant. If there are several reactants this becomes:

$$\mathrm{d}c/\mathrm{d}t = -k\,c^1 \,.\,.\,.\,.\,.\, c^2 \,.\,.\,.\,.\,.\, c^n.$$

The order of the reaction is how the rate of the reaction depends on the concentration of the reactants. Thus:

$$2\,NO_2 \rightarrow 2\,NO + O_2$$
$$\mathrm{d}(NO_2) = k(NO_2)^2$$

is a second-order reaction. The conversion of *kerogen to oil is assumed to be a *first-order reaction. *See also*: chemical kinetics. *References*: Physical chemistry textbooks.

organic facies A distinctive assemblage of *kerogen components which can be visually identified, or has a characteristic chemical composition, is called an organic facies. The organic facies is controlled by primary organic matter input, preservational conditions, and energy levels in the depositional environment; it may cross lithostratigraphic and chronostratigraphic boundaries. Organic facies can be predicted from depositional models and is of importance in mapping the distribution of, and compositional variations in, *source rocks. *See also*: palynofacies.

organic geochemistry The study of the visual and chemical composition of *organic matter in unconsolidated sediments and rocks, and of the changes induced in it by bacterial and thermal alteration with time.

organic matter Biogenic constituents of sedimentary rocks, also termed sedimentary organic matter (SOM). It is composed of insoluble *kerogen and soluble *bitumen or oil. Different types of organic matter give different products in response to increases in temperature over geological time. They may generate carbon dioxide, water and oil, and/or *hydrocarbon gases.

organic petrology The microscopic study of sedimentary organic matter using transmitted and reflected white and UV light. *See also*: coal petrography, maceral. *Reference*: Teichmuller, M. (1985). Organic petrology of source rocks, history and state of the art. *Org. Geochem.*, **10**, 581–99.

organic solvents Liquids which are used to dissolve oil and oil-like components from rocks, and also to subdivide oils and extracts into chemical fractions. Examples are pentane, dichloromethane, *benzene, and methanol.

organometallic compounds Organic compounds which contain metals complexed into their structure are called organometallic compounds. The most significant organometallic compounds in *organic geochemistry are the *porphyrins. The most important metals found in oils and sediments as porphyrins are nickel and vanadium, which can reach concentrations of 1000 p.p.m. Other metals which can be found in organic matter in sediments and oils are iron, copper, lead, arsenic, molybdenum, magnesium, cobalt, manganese, and chromium.

oxic *See* aerobic.

oxygen The gaseous element, O, atomic weight 16. It has three isotopes ^{16}O, ^{17}O, and ^{18}O. Oxygen occurs as a minor component of most oils in the form of phenols, ketones, acids, and fluorenones (see Fig.), which are part of the *polar or *NSO fraction of oils.

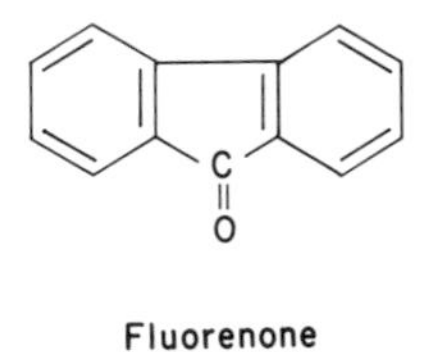

Fluorenone

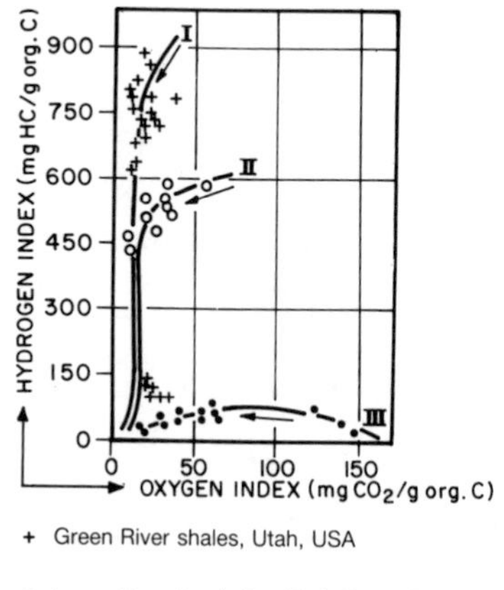

oxygen index A parameter which measures the oxygen richness, usually of *kerogen. It is derived from *Rock Eval and related types of *pyrolysis. It is calculated from the S_3 and *total organic carbon (S_3/TOC). It has a direct relationship with elemental O/C ratios (see Fig.), and is measured in units of mg CO_2/g organic carbon. In conjunction with the *hydrogen index it may be used to determine *kerogen type and level of *maturation. The index is unreliable in host rocks which are high in carbonates, so removal of carbonate by acid is needed prior to analysis. Several reasons for this have been discussed, including release of occluded carbon dioxide from the carbonate matrix on heating or minor low temperature breakdown of carbonates. High values >50 mg/g are characteristic of *immature kerogen (see Fig.). *See also*: hydrogen index.

oxygen isotopes Oxygen has three *stable isotopes, ^{16}O, ^{17}O, and ^{18}O. ^{18}O and ^{16}O are used to determine *palaeotemperatures. The oxygen isotope standard was originally *PDB but is now *SMOW. The conversion of SMOW to PDB is by the formula:

$$\delta^{18}O(SMOW) = 1.03086\ \delta^{18}O(PDB) + 30.86$$

Other standards used are NBS-18 and NBS-19.

ozocerite *See* solidified bitumen.

P_0 P_1 P_2 P_3 P_4 The *Rock Eval and *Oil Show Analyzer peaks whose areas are S_0 S_1 S_2 S_3 and S_4 respectively. They equate to the thermal extraction of existing gases (P_0) and oils (P_1), the *pyrolysis of *kerogen (P_2), the total CO_2 emitted during pyrolysis (P_3) and the residual carbon (P_4). *See also*: Oil Show Analyzer, Rock Eval, S_0 S_1 S_2 S_3 S_4.

palaeotemperature analysis 1. The analysis of water temperatures during carbonate or phosphate deposition by $^{16}O/^{18}O$ *isotope or unsaturated ketone analysis.

2. The analysis of the maximum temperature (maturity) of source rocks by examination of *vitrinite reflectance, kerogen colour, or chemical composition. It is sometimes used specifically to describe ESR results, called maximum palaeotemperature analysis. *See also*: maturation.

palynofacies An assemblage of recognizable palynomorphs and dispersed *organic matter, preserved in sedimentary rocks, which is used to develop predictive depositional models. The final assemblage of organic matter in rocks is dependent on three factors; primary organic matter input, preservation, and depositional processes. Primary organic input is dominantly controlled by age, climate, nutrient supply, and water salinity. Factors controlling preservation are Eh, pH, and temperature. Energy levels, current direction, and velocity control the distribution of the surviving organic matter in the same way as they control the distribution of mineral components of rocks.

Palynofacies can be used in addition to normal sedimentological interpretation in sands and silts, but has the advantage of defining depositional environments in shales too. Massive sands may also yield facies-specific organic matter assemblages. Another use of palynofacies analysis is in the prediction of the occurrence of and facies changes in *source rocks. *See also*: organic facies. *Reference*: Hancock, N.J. and Fisher, M.J. (1981). Middle Jurassic North Sea deltas with particular reference to Yorkshire. In *Petroleum Geology of the Continental Shelf of North-west Europe*, (ed. L.V. Illing and G.D. Hobson). Institute of Petroleum, Heyden, London.

palynomorph A recognizable organic fossil derived from plants and animals including microforaminiferal linings, scolecodonts, graptolite siculae, spores, pollen, fungae, acritarchs, dinocysts, and marine and non-marine *algae.

paraffin *See* normal alkane.

parent ion The lower energy *'molecular ion', whose decay to a *daughter ion is observed during metastable transitions. By selecting the relevant parent ion the transitions of a well-defined group of isomeric compounds can be observed, usually to the *diagnostic ion for the type of compound under study. A double focusing mass spectrometer (*GC–MS–MS system) is required to observe these transitions. *See also*: metastable ions.

PDB The Chicago PeeDee belemnite. The PeeDee Formation of Cretaceous age (USA), contained the PeeDee belemnite, which was used as the first standard for *carbon isotope and *oxygen isotope analysis. The original standard material has long since been used up, but other standard materials have been calibrated against PDB and it is still conventional to express carbon isotope ratios relative to PDB. The conversion factor, *SMOW, to PDB for oxygen is:

$$\delta^{18}O(SMOW) = 1.03086 \times \delta^{18}O(PDB) + 30.86$$

and for PDB to *NBS-22 for carbon is:

$$\delta^{13}C(PDB) = 0.9702 \times \delta^{13}C(NBS\text{-}22) - 29.8$$

peak mature The level of maturity when the maximum rate of conversion of *kerogen to oil or gas occurs. For *oil prone *source rocks this equates to a *spore colour index (scale 1 to 10) of 5.0 to 7.0, *thermal alteration index (Staplin) of 2.3 to 2.6, and *vitrinite reflectance of 0.65 to 0.9 per cent, and for *gas prone source rocks a TAI of 2.6 to 3.6 and vitrinite reflectance 1.3 to 2.2 per cent. *See*, Table 1, summary section, for maturation summary. *See also*: early mature, late mature, post mature, immature.

peat Peat is the natural product of the decay of higher plants in terrestrial *anaerobic environments. The initial stages of *coal formation are similar to peat although not identical, as climate and plant types are different. A typical peat has the composition, 5 per cent water-soluble components, 45 per cent *cellulose, 10 per cent soluble organic matter, 23 per cent *lignin, 2 to 3 per cent others.

petroporphyrins *See* porphyrins.

phenanthrene A simple *aromatic compound containing three *benzene rings (see Fig.). The relative abundance of monomethyl-substituted phenanthrenes, which occur at the 1-, 2-, 3-, and 9- positions, is thermodynamically controlled, and this forms the basis of the *methyl-phenanthrene index or MPI. *See also*: Methylphenanthrene Index.

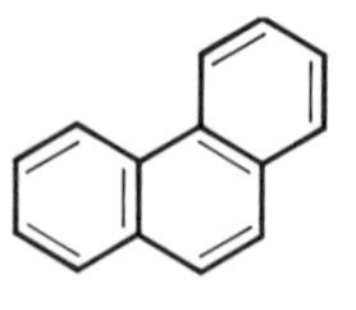

Phenanthrene

photic zone *See* euphotic zone.

photosynthesis The process whereby carbon dioxide and water are converted to *cellulose using energy from sunlight. The reaction is catalysed by *chlorophyll. A simple equation describing the process is:

$$6CO_2 + 6H_2O \leftrightarrow C_6H_{12}O_6 + 6O_2$$

phytane The C_{20} *regular isoprenoid *hydrocarbon formula $C_{20}H_{42}$, also known as 2,6,10,14-tetramethylhexadecane. It is ubiquitous in most oils and sediment extracts, and is a major biological marker. It is mainly derived from the side chain of the *chlorophyll molecule, although it has also been found in *methanogenic bacteria. In reducing environments it is formed by hydrogenation of *phytol to dihydrophytol, and then to phytane (see Fig.). In *gas chromatography, it is distinctive as a doublet with C_{18} *normal alkane, in the *saturate fraction of oils and sediment *extracts. Its abundance relative to n-C_{18} may be a maturation parameter, and can also indicate *biodegradation. Isoprenoids are destroyed during intense biodegradation, generally following the removal of normal alkanes. The abundance of phytane relative to *pristane is used in *oil–source correlations. Some oils of Ordovician age have been found to be devoid of both pristane and phytane; they are derived from a particular alga which did not use chlorophyll. *See also*: phytane/n-C_{18} ratio, pristane/phytane ratio.

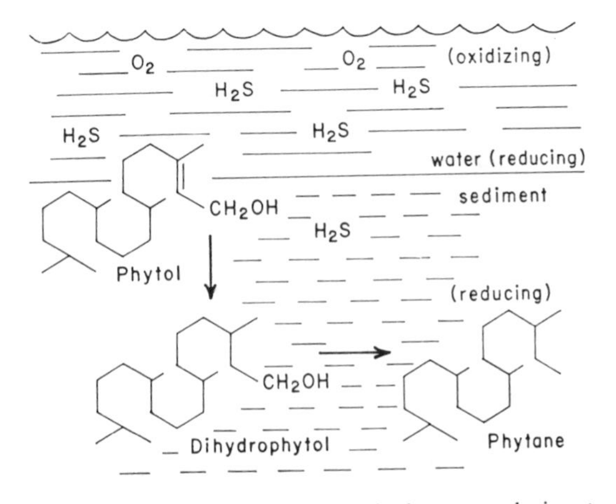

phytane/n-C_{18} ratio The abundance of phytane relative to n-C_{18} as determined from *gas chromatography of oils and sediment *extracts. The ratio may be used as an environmental parameter, and to indicate maturity or *biodegradation. Values in oils and *mature source rocks are in the order of 0.5. In moderately biodegraded oils and *immature source rocks, phytane is more abundant than n-C_{18} so values are generally greater than 1.0; in very mature oils and source rocks, the value may be as low as 0.10.

phytol The *diterpenoid primary product from the decomposition of the *chlorophyll molecule in shallow sediments, formula $C_{20}H_{39}OH$

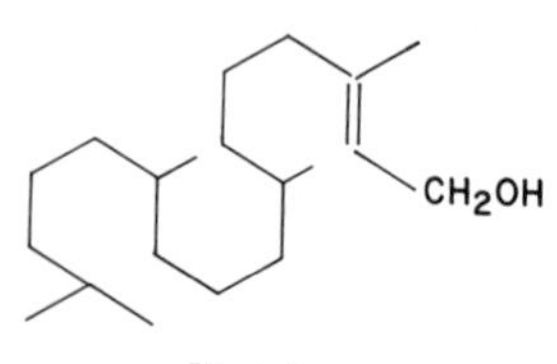

Phytol

(see Fig.). The phytol double bond is hydrogenated to dihydrophytol in reducing conditions and, in more oxidizing conditions, it is oxidized to phytenic acid. These two diagenetic pathways lead to *phytane and *pristane. The relative abundance of these two compounds is hence a measure of environmental conditions during deposition and early diagenesis. *See also*: phytane, pristane.

piezometric mapping *See* potentiometric mapping.

pigments The colour of organic materials is due to the presence of pigments. These can be preserved relatively unaltered in sediments (e.g. β-carotene as β-carotane), or they may be decomposed, e.g. haemin or *chlorophyll to *porphyrins, *pristane, and *phytane. These compounds are *biological markers. *See also*: carotenoids.

polar fraction That fraction of *crude oils and sediment *extracts which contains the *polar compounds. It is usually eluted with methanol during *liquid chromatography. Synonymous with *NSO, *hetero compounds, and *resins. *See also*: NSO fraction.

porphyrins Organometallic derivatives of the nucleus of the haemin or *chlorophyll molecule. Although iron and magnesium are the metals bonded to these compounds as natural products, during *diagenesis ion exchange takes place, and nickel and vanadium are substituted. The relative abundance of Ni to V is controlled by water and sediment redox potential, and pH. Although the metals in porphyrins are usually Ni and V, other metals are also found, e.g. copper and iron. They are heteroatomic molecules composed of C, H, and N. Four basic types of alkyl porphyrins are recognized in sedimentary organic matter—well-documented and abundant de(*s*)oxophylloerythro(*a*)etioporphyrin (*DPEP), containing an exocyclic alkano ring, and found in chlorophylls (see Fig., p. 99); well-documented and abundant etio porphyrins with no exocyclic alkano ring (see Fig., p. 99); rhodo porphyrins, with a tetrapyrrole structure; di-DPEP with one more degree of unsaturation compared with DPEP. The relative concentrations of DPEP to etio decreases by cracking. Their relative abundances have been used as a contentious *maturation parameter. They are part of the *polar fraction of oils and are separated by *liquid chromatography then analysed by *gas chromatography–mass spectrometry. Synonymous with petroporphyrins. *Reference*: Baker, E.W. and Louda, J.W. (1986). Porphyrins in the

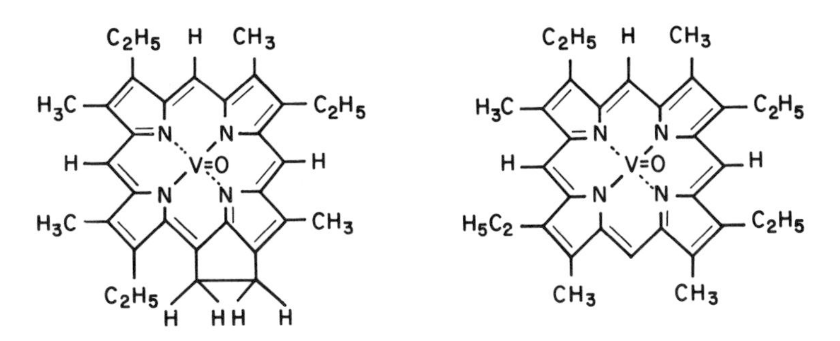

geologic record. In *Biological Markers* (ed. R.B. Johns), pp. 125–225. Elsevier, Amsterdam.

positional isomer *See* isomer.

post mature The level of *maturation beyond the main phase of *hydrocarbon generation. It merges into early inorganic metamorphism (greenschist facies), and is equivalent to metagenesis. *Oil prone *source rocks are post mature at *spore colour index >9.0 (scale of 1 to 10), *thermal alteration index (Staplin) >3.5, *vitrinite reflectance >1.3 per cent; *gas prone source rocks are post mature at TAI >4.0 and vitrinite reflectance >3.0 per cent. *See*, Table 1, summary section, for maturation summary. *See also*: immature, peak mature, late mature, early mature.

potential source rock A source rock which has not yet generated significant amounts of *hydrocarbons due to immaturity. *See also*: effective source rock.

potential yield The total amount of already generated and potential *hydrocarbons from a *source rock, as determined from *pyrolysis, e.g. $S_1 + S_2$ from *Rock Eval or $S_0 + S_1 + S_2$ from an *Oil Show Analyzer. Values may be quoted in p.p.m., kg/tonne, barrels/acre ft. Inaccuracies may be caused by contamination, e.g. migrated hydrocarbons, or drilling mud additives.

potentiometric maps Maps developed by hydrogeologists to study the movement of fluid in the subsurface. Pressure and formation depth information are plotted to chart the distribution of aquifers. These maps have been adopted by petroleum geologists and geochemists to project fluid flow patterns for *secondary migration route mapping. Oil moves in response to the forces of buoyancy and pressure, so that its route from *source rock to reservoir can be delineated (see Fig., p. 100). Synonymous with piezometric maps, water table maps.

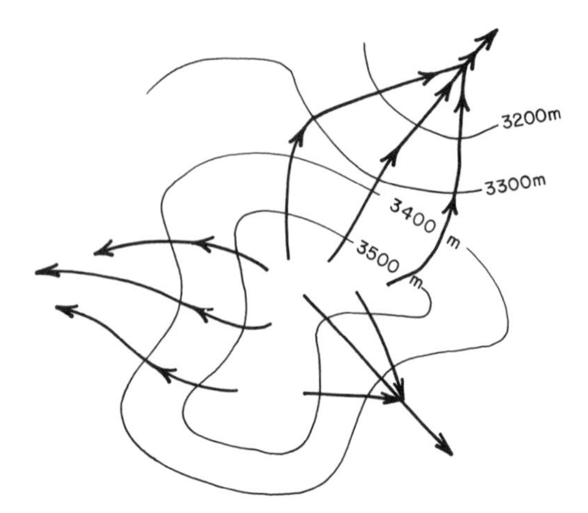

pre-exponential factor *See* A factor.

primary migration The migration of oil from within its fine grained source rock into a more porous, permeable *carrier rock is called primary migration (see Fig.). The exact mechanism of primary migration has not been proven, but modelling of rock properties has suggested that internal pressure builds up due to inorganic and organic transformations, leading to formation of a network of microfractures, and eventual *expulsion of oil and gas. Other suggested mechanisms such as solution in water and diffusion are less favoured.

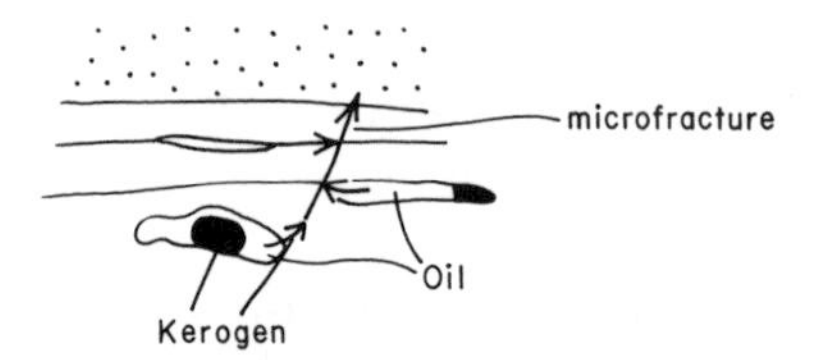

Primary migration

Primary migration is an inefficient process and does not act equally on all of the chemical components of oils, so that compositional changes have been observed between *source rock extracts and expelled oils. *Normal alkanes are more efficiently expelled than *branched alkanes. Primary migration efficiencies of between 5 and 80 per cent of available oil are believed to be appropriate for most source rocks, depending on thickness, lithology, *kerogen type and quantity, and level of maturity. Primary migration of oil is less efficient than of gas. Synonymous with expulsion. *References*: Palciauskas, V.V. and Domenico, P.A. (1980). Microfracture development in compacting sediments: relation to hydro-

carbon maturation kinetics. *Am. Assoc. Pet. Geol. Bull.*, **64**, 927–37.
Leythaeuser, D., Mackenzie, A.S., and Schaefer, R.G. (1984). A novel approach to the recognition and quantification of hydrocarbon migration effects in shale-sandstone sequences. *Am. Assoc. Pet. Geol. Bull.*, **68**, 196–219.

pristane The C_{19} *regular isoprenoid hydrocarbon, formula $C_{19}H_{40}$, derived from the side chain of the *chlorophyll molecule; it is ubiquitous in most oils and sediment *extracts: also known as 2,6,10,14,-tetramethylpentadecane. It is one of the most important *biological markers. It is used for *oil to oil and *oil–source rock correlations.

Pristane is usually derived from chlorophyll via phytol in oxygenated environments, where the *phytol is converted to phytenic acid before *decarboxylation to pristene, and hydrogenation to pristane (see Fig.). In *gas chromatography it occurs as a distinctive doublet with normal C_{17} alkane in oils and sediment extracts. Its abundance relative to n–C_{17} is an environmental, *maturation and *biodegradation indicator. It may be removed by moderate to severe biodegradation. Its abundance relative to phytane is also an environmental indicator. It has been found in tocopherols, but found to be absent in some Ordovician oils, and very highly mature oils from most sources. *See also*: pristane/n-C_{17} ratio, pristane/phytane ratio.

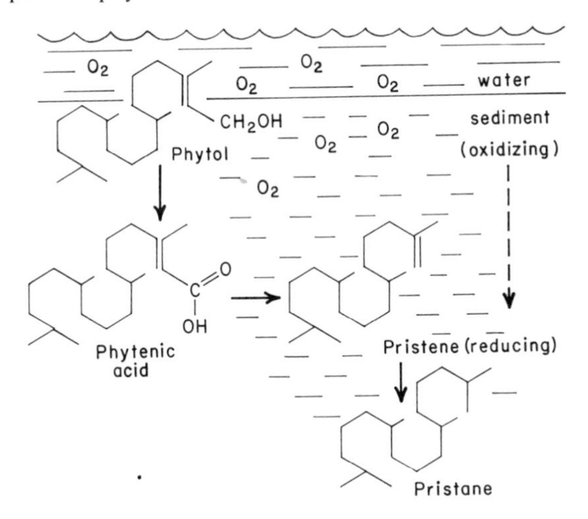

pristane/n-C_{17} ratio The relative abundance of *pristane relative to n-C_{17} (the main algal alkane) as determined from *gas chromatography of oils and sediment extracts, gives a ratio used as an environmental parameter. It is also used to indicate *maturation, *migration, and *biodegradation. Values in oils and relatively *mature source rocks are in the order of 0.5, but may be as low as 0.1. In moderately biodegraded oils and *immature source rocks, pristane is more abundant than n-C_{17}, so that values may be greater than 1.0. *Reference*: Leythaeuser, D. and Schwarzkopf, T. (1986). The pristane/n-heptadecane ratio as an indicator for recognition of hydrocarbon migration effects. *Org. Geochem.*, **10**, 191–7.

pristane/phytane ratio The ratio of the relative abundance of the *isoprenoids *pristane to *phytane determined from *gas chromatography. Both of these compounds can be derived from the *chlorophyll molecule under different Eh conditions, more pristane than phytane being produced under more oxidizing conditions. A high pristane/phytane ratio (>1), indicates more oxidation; a low pristane/phytane ratio (<1) indicates more reducing conditions. The ratio is also used as an *oil to oil and *oil–source correlation parameter. As the ratio can be affected by many factors it is not used as a major correlation parameter. Marine carbonate sources tend to have a pristane/phytane ratio of <1. *Reference*: Dydik, B.M., Simoneit, B.R.T., Brassell, S.C., and Eglinton, G. (1978). Organic geochemical indicators of palaeoenvironmental conditions of sedimentation. *Nature*, **272**, 216–22.

Production Index, P.I. A maturity parameter derived from *pyrolysis, especially *Rock Eval and *Oil Show Analyzer, which is the ratio of already generated *hydrocarbons to potential hydrocarbon. The ratio is calculated, for gas, from the formula:

$$S_0/S_0 + S_1 + S_2$$

and, for oil, from:

$$S_1/S_0 + S_1 + S_2 \text{ (OSA) and } S_1/S_1 + S_2 \text{ (Rock Eval).}$$

*Immature samples have a ratio of 0.1 or less, *mature samples, 0.1 to 0.4. When *expulsion occurs, the S_1 no longer increases. The presence of migrated oil affects the ratio.

productivity 1. The primary productivity of biological material has varied throughout geological time because of evolution, distribution of continental shelf and the oceans, and global temperature changes. The type and quantity of *organic matter in sediments is related to this primary productivity and preservation. In the Precambrian to Late Devonian, *bacteria, *algae, and zooplankton were dominant. In the Devonian to Jurassic, higher land plants were first seen, but they did not become dominant until the Cretaceous to Recent. Oceanic productivity is high where nutrient supply is good, such as in colder water and in areas of upwelling. High oceanic productivity is considered to be 600 t/m^2 y. Open oceans in equatorial regions are areas of low productivity, (50 t/m^2 y). The present worldwide annual *organic matter production is estimated to be 6×10^{10} metric tonnes.

2. The productivity of a *source rock is defined as the amount of

*hydrocarbons which can theoretically be generated from its *kerogen. This is calculated from *pyrolysis results. The units are barrels/acre ft., or kg/tonne, or m^3 oil/km^3 rock. This can be calculated from *Rock Eval S_1 and S_2 data directly (dividing by 36 to convert from p.p.m. to bbls/acre ft.). The present productivity of a source rock can be calculated from Rock Eval S_1 or *extract data in the same way. Total conversion of kerogen to hydrocarbons is unrealistic, so that, for volumetric estimates of hydrocarbon generation, efficiency factors must be applied for conversion of kerogen, as well as for *migration efficiency.

propane The gaseous *normal alkane containing three *carbon atoms, formula C_3H_8 (see Fig.). It is a ubiquitous major component of all natural oil and gas accumulations, except *biogenic gas.

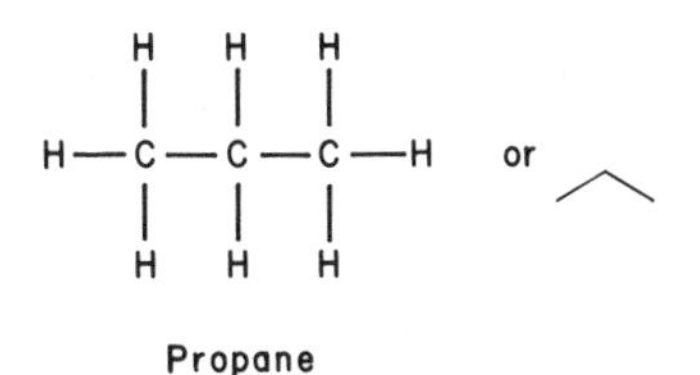

Propane

proteins Proteins form the bulk of lower plants and animals, and are minor components of higher plants. They form enzymes and other diverse components of *organic matter. They are made up of complex polymers of L-*amino acids, and are hence of high molecular weight. They are insoluble in water but decompose readily to amino acids in the water column and surface sediments. *Reference*: Hare, P.E. (1969). Geochemistry of proteins, peptides and amino acids. In *Organic Geochemistry* (ed. G. Eglinton and M.J. Murphy), 828 pp. Springer-Verlag, Berlin.

pseudoactivation energy The apparent *activation energy, especially of *kerogen, measured from *pyrolysis experiments, e.g. *Rock Eval. It is not a true measure of activation energy as it may reflect other processes (such as diffusion or removal of steric hindrance) as well as kinetic factors; the term 'pseudoactivation energy' is hence used. There is a difference between predicted and measured activation energies, which was originally attributed to catalysis. The discrepancy is now thought to be due to several reactions occurring in parallel with true activation energies 45 to 55 kcals, the effect being to simulate a reaction with a much lower activation energy (20 kcals). *Reference*: Waples, D.W. (1984). Thermal models for oil generation. In *Advances in Petroleum Geochemistry* I (ed. J. Brooks and D. Welte). Academic Press, London.

PY–GC *See* pyrolysis–gas chromatography.

pyrite Cubic mineral containing iron and *sulphur, known as fools gold. It can occur as authigenic cubes in sediments or, if associated with *organic matter, in framboidal form, where it indicates synsedimentary activity by *anaerobic bacteria. It is difficult to separate from *kerogen in rocks, and can be confused with *inertinite in *late mature kerogen in transmitted white light. *Reference*: Raiswell, R. and Berner, R.A. (1985). Pyrite formation in euxinic and semi-euxinic sediments. *Am. Jour. Sci.*, **285**, 710–24.

pyrobitumen The insoluble residue which remains in *source rocks after the generation of oil and gas, formed by the *in situ* cracking of oil which has not migrated from the source rock. The products of this cracking are an insoluble residue, pyrobitumen, and gas. It may be difficult to distinguish from primary *kerogen. *See also*: solidified bitumen.

pyrofusinite *Fusinite formed from the burning of plant matter, of distinctive morphology, as cell lumens are often preserved. *See also*: fusinite.

pyrolysis The process of heating a rock or *kerogen sample in the laboratory to generate *hydrocarbons by thermal decomposition is called pyrolysis. The heating can be either in the presence of water, when it is called *hydrous pyrolysis or in the absence of water, when it is called anhydrous pyrolysis. *Rock Eval is anhydrous pyrolysis. It is normally carried out to simulate *maturation, or to measure *activation energy. *See also*: hydrous pyrolysis, Rock Eval. *Reference*: Harwood, R.J. (1977). Oil and gas generation by laboratory pyrolysis of kerogen. *Am. Assoc. Petr. Geol. Bull.*, **61**, 2082–101.

pyrolysis–gas chromatography An analytical technique which allows the products of *pyrolysis to be examined by *gas chromatography (GC). The actual chemical composition of, for example, *Rock Eval P_1 and P_2 peaks can be examined. The results are similar in appearance to the GC analysis of sediment *extracts. The technique provides information on likely *hydrocarbon products from source rocks on increasing *maturation. *See also*: gas chromatography–mass spectrometry. *Reference*: Dembicki, H., Horsfield, B., and Ho, T. T.Y. (1983). Source rock evaluation by pyrolysis–gas chromatography. *Am. Assoc. Petr. Geol. Bull.*, **67**, 1094–103.

quality of a source rock A description of the potential of a source rock to generate oil or gas based on the quantity of organic carbon and its *kerogen type.

quantity of organic carbon A measure of the weight per cent of organic carbon in a rock, also known as the *Total Organic Carbon or TOC. It may be determined chemically or pyrolytically. *Source rocks are described as lean or poor at <0.5 per cent, fair at 0.5 to 1.0 per cent, moderate at 1.0 to 2.0 per cent, good at 2.0 to 5.0 per cent, and very good, rich, or excellent at >5.0 per cent. The average shale contains 0.9 per cent TOC. New compilations of data suggest that the average source rock contains >2 per cent TOC. *Reference*: Tissot, B. and Welte, D. (1984). *Petroleum formation and occurrence* (2nd edn), 699 pp. Springer-Verlag, Berlin.

***R* (rectus)** The term used to describe an *optical isomer according to the Cahn–Ingold–Prelog system, which replaces D or dextrorotatory, with clockwise (+) rotation of polarized light. The system is based on a complex hierarchy of substituent group atomic number, bond type, and spatial orientation. Each *chiral centre is numbered separately, e.g. C-20 *R* steranes. *References*: Organic chemistry textbooks.

R_o Common abbreviation for *vitrinite reflectance under oil immersion.

racemic A mixture which contains equal amounts of two *enantiomers and hence has no overall *optical activity. *References*: Organic chemistry textbooks.

rank *See* coal rank.

rate constant The proportionality coefficient in the rate equation, determining the dependence of the reaction velocity on concentration of the reactants. The basic rate equation is a first-order differential equation in the form:

$$-\mathrm{d}c/\mathrm{d}t \propto c,$$

where c is the concentration of reactants and t is time. It follows that:

$$-\mathrm{d}c/\mathrm{d}t = kc,$$

where k is the rate constant. Synonymous with specific rate of a reaction. *References*: Physical chemistry textbooks.

rate of a reaction The quantity of a reactant that is converted into product during a specified time period gives the rate of reaction. Among the factors which affect reaction rates are temperature, catalysts, concentration, and nature of the reactants. Reaction rates tend to increase with concentration and temperature. *See also*: order of a chemical reaction. *References*: Physical chemistry textbooks.

rearranged steranes *Biological marker compounds derived from plant and animal *steroids. They range from *carbon number C_{27} to C_{29}

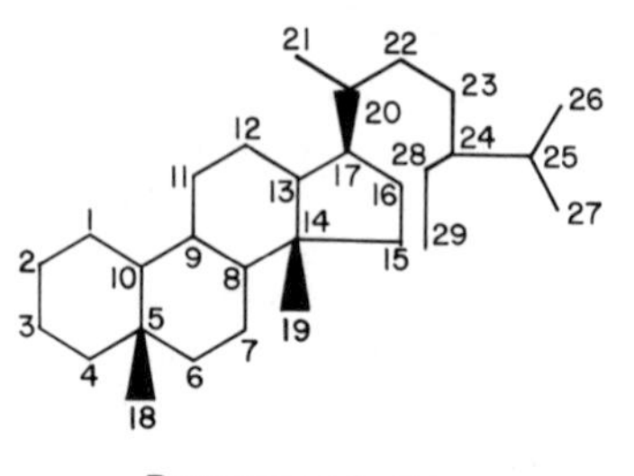

Rearranged steranes

(see Fig.). They are formed by acid-catalysed 'backbone' rearrangement of steroids, so that their absence indicates a carbonate or evaporitic *source rock (the absence of clays that promote acid catalysis). Important *optical isomers are the C-20 *R and C-20 *S forms, and important *stereoisomers are the 13(H), 17(H) α and β isomers. 13α(H), 17β(H) isomers are more commonly encountered in *immature source rocks, but 13β(H), 17α(H) is the more thermally stable *isomer. They are detected during *gas chromatography–mass spectrometry analysis of the *saturate fraction of oils and sediment *extracts at m/e of 217, 232, and 259. These rearranged steranes are more resistant to bacterial attack than *regular steranes. Within the rearranged steranes, C_{29} steranes are more resistant to *biodegradation than C_{27} steranes. The relative proportion of rearranged steranes to total steranes increases with increasing maturity. Synonymous with diasteranes.

refinery analysis The analysis of *crude oils and *condensates which provides information on the way that they will behave during refining. The analysis includes pour point, flash point, wax content, Ni, V, and S contents, boiling point analysis, and distillation analysis etc.

regular isoprenoids *Isoprenoids formed by 'head to tail' linkages of the *isoprene unit, e.g. *pristane. *See also*: isoprene.

regular steranes C_{27} to C_{30} *steranes with 5(H), 14(H), 17(H) α- or β-stereochemistry (see Fig.). *Optical isomers (*R and *S forms) occur at the C-20 and C-24 positions, but most conventional *gas chromato-

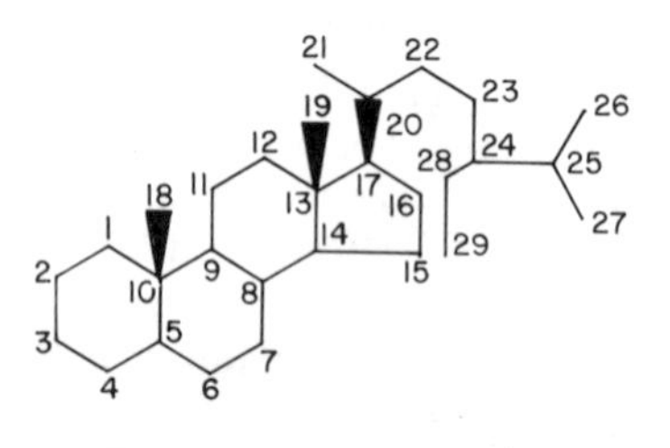

Regular steranes

graphy–mass spectrometry (GC–MS) systems are unable to separate the optical isomers at the C-24 position. They are detected by GC–MS analysis of the *saturate fraction of oils and *extracts at *m/e* 217 and 218.

The relative abundance of $C_{27}:C_{28}:C_{29}$ parent sterols was found to be environmentally related. This relationship has been directly transferred to steranes to interpret environment of deposition of the organic matter from sediments and oils. The abundance of C_{29} steranes was originally assigned to the influence of higher land plants, but their abundance in sediments before the evolution of land plants indicates that they must have more than one source. The relative abundance of $C_{27}:C_{28}:C_{29}$ steranes is now used more cautiously. The abundance of C_{30} steranes, minor components of oils and sediment extracts, is believed to indicate a derivation from marine sources.

The relative abundances of different sterane *isomers can be used as a measure of maturity of an oil or sediment. The early *diagenetic sterane isomer abundance is 80 per cent $\alpha\alpha\alpha$, 20 per cent $\beta\alpha\alpha$. With increasing *maturation, both of these isomers convert to the $\alpha\beta\beta$ stereochemistry. The relative abundance of $\alpha\beta\beta/\alpha\alpha\alpha$ steranes is used as a maturity parameter, usually for the C_{29} steranes (there are co-elution problems with C_{27} and C_{28} regular steranes with normal GC–MS. If *metastable ions are used, then these parameters can be used for steranes of all carbon numbers). Regular steranes disappear during intense *biodegradation, and they are removed in the order $C_{27}>C_{28}>C_{29}$; the biologically inherited $14\alpha(H)$, $17\alpha(H)$ **R* isomer is more readily removed than the **S* form.

residual oil Heavy, *asphaltene-rich, immovable oil found at oil–water contacts, on migration routes, and in reservoirs. It indicates the previous residence of oil, and may suggest that seal failure, *biodegradation, gas flushing, or tectonic movement may have disturbed the oil. *See also*: tar mat.

resinite A term used in the microscopic examination of *coals to describe the fossilized remains of plant resins which bleed from damaged or cut bark surfaces, waxes, and essential oils (*terpenoids). It is durable, but does not travel far from the site of origin. It may develop either as small, separate, commonly oval bodies, or as inclusions in *telinite and *collinite. It may flow at high temperatures and fill fractures; it is sometimes erroneously called *exsudatinite. It is lemon-yellow to yellow-orange with red internal reflections, in transmitted white light, and grey to dark grey in reflected light. Resinite may show anisotropy, and commonly shows oxidation rims. Low rank resinite fluoresces strongly from blue-green to pale orange in UV light.

Quantities of resinite are found in a belt from South America to New Guinea, related to the palaeogeography before continental break-up about 170 million years ago. It is found in the Cretaceous to Miocene of

New Zealand, and in significant quantities in the Yallourn brown coal of Australia. Coals which contain resinite are usually *cannel coals. For some time it has been postulated that resinite can source oils. A review of literature suggests that resinites are hydrogen-rich, but do not source significant quantities of liquids. The *H/C ratio varies from 1.0 to 1.5. Synonymous with amber. It equates chemically with *Type II kerogen. *Reference*: Mukhopadhyay, R.K. and Gormly, J.R. (1984). Hydrocarbon potential of two types of resinite. *Org. Geochem.*, **6**, 394–454.

resins The fraction of *crude oil or sediment *extracts which elutes with methanol during *liquid chromatography. It contains the high molecular weight *NSO compounds and *asphaltenes (if not removed prior to liquid chromatography). The term is used by some analysts to include lower molecular weight NSO compounds. Synonymous with *polar, *NSO fraction, and *hetero compounds. *See also*: column chromatography. *Reference*: Pelet, R., Behar, F., and Monin, J.C. (1986). Resins and asphaltenes in the generation and migration of petroleum. *Org. Geochem.*, **10**, 481–98.

retrograde condensates When the pressure in a gas *condensate accumulation is reduced by production, condensation may occur in the reservoir, and in the well bore. This is known as retrograde condensation. This is the reverse of normal condensate production and is detrimental to the production characteristics of the accumulation, as the condensate is not recoverable. Retrograde condensation may be prevented by recycling the gas already produced, after separation from the condensate at the surface.

reworked kerogen Recycled *kerogen particles derived from the erosion of older rocks. These may be redeposited in large quantities in younger sediments if the older rocks are rich in *organic matter, e.g. coals. Reworked spores and *vitrinite are often found. Spores can be fairly easy to identify if they are age-diagnostic. Reworked vitrinite is often impossible to distinguish from *in situ* vitrinite as there are no absolutely diagnostic morphological characteristics, but it can sometimes be distinguished by being slightly more rounded or blocky, having oxidation rims and higher relief. Sometimes it can show pits where *pyrite has been weathered out, but reworked vitrinite may also contain pyrite. Reworked kerogen is often assumed to have a lower hydrocarbon potential than *in situ* kerogen, as it may be oxidized during the recycling. It may only be identified during *vitrinite reflectance analysis, when it is of higher maturity than the *in situ* material. Synonymous with *allochthonous.

rift basin models Thermal models used to explain the origin and development of sedimentary basins in tensional tectonic regimes. They are of interest to *organic geochemistry as they enable *palaeotemperature to be predicted in *maturation modelling. The most widely used model for the prediction of *heatflow in rift basins is that of McKenzie. *Reference*: McKenzie, D.P. (1978). Some remarks on the development of sedimentary basins. *Earth and Planet. Sci. Lett.*, **40**, 25–32.

Rock Eval A commercial technique for the anhydrous *pyrolysis of *source rocks developed by the Institute Français du Pétrole (IFP); it enables the chemical composition of *kerogen, and hence its hydrocarbon potential, to be determined. There are two automatic pyrolysis steps, the first being an initial volatilization of pre-existing hydrocarbons in the source rock at 300 °C. This is done in a stream of helium gas, and produces the P_1 peak of area S_1. The second step is the conversion of kerogen to hydrocarbons by increasing the sample temperature to 550 °C, again in an inert helium atmosphere. This produces the P_2 peak of area S_2. Carbon dioxide produced during the pyrolysis up to a temperature of 390 °C, is collected and is the P_3 peak of area S_3 (see Fig.). *Maturation and source quality parameters derived from the peaks are: *hydrogen index (S_2/*total organic carbon, TOC); *oxygen index (S_3/TOC); *production index ($S_1/S_1 + S_2$); *potential yield ($S_1 + S_2$ expressed as kg oil/tonne rock or barrels oil per acre ft. rock), and *T_{max} from the S_2 peak. The hydrogen index is a kerogen typing parameter, and the oxygen index and T_{max} mostly used as maturation parameters, but they are also affected by kerogen type. Results are adversely affected by drilling-mud contamination, *mineral matrix effects, and low amounts of kerogen. They are obtained quickly and inexpensively, and the technique is an excellent way of monitoring samples for further types of geochemical investigation. *See also*: hydrogen index, mineral matrix effects, oxygen index, potential yield,

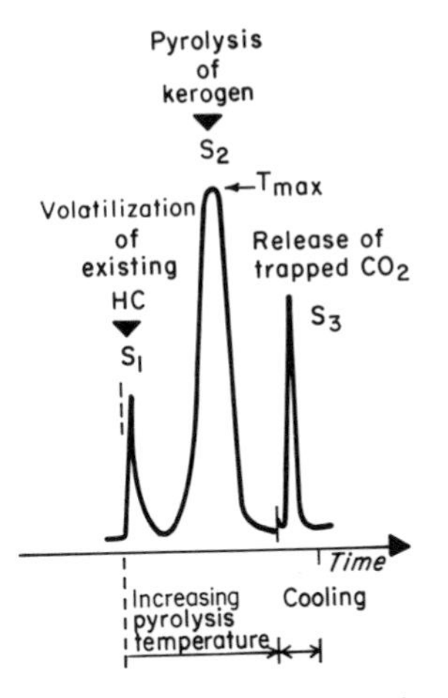

A typical Rock Eval output

Production Index. *Reference*: Orr, W.L. (1983). Comments on pyrolytic hydrocarbon yields in source rock evaluation. In *Advances in Organic Geochemistry 1981* (ed. M. Bjorøy *et al.*). John Wiley, Chichester.

***S* sinister** The term used to describe an *optical isomer according to the Cahn–Ingold–Prelog system, which replaces L or laevorotatory, with anticlockwise (−) rotation of polarized light. The system is based on a complex hierarchy of substituent group atomic number, bond type, and spatial orientation. Each *chiral centre is numbered separately, e.g. C-20 *S* steranes. *References*: Organic chemistry textbooks.

S_0 S_1 S_2 S_3 S_4 *Rock Eval and *Oil Show Analyzer pyrolysis parameters derived from the areas of the P_0, P_1, P_2, P_3, and P_4 peaks respectively (see Fig.). S_0 is the already generated gases and S_1 the already generated oil, measured in units of p.p.m. (S_0 is only available from the Oil Show Analyzer); these values are badly affected by migrated *hydrocarbons and contamination. S_2 is the remaining hydrocarbon potential, measured in p.p.m. It too can be affected by high molecular weight contamination, and by *mineral matrix effects. S_3 is the measure of the carbon dioxide released during pyrolysis, and is proportional to the *oxygen present in the *kerogen and measured in p.p.m.; it is only available from the Rock Eval. The S_3 can be increased in carbonate rocks by processes which are not fully understood but which may be from CO_2 occluded in the carbonate, or by low-temperature carbonate decomposition. The S_4 is equivalent to residual carbon, and also only determined by the Oil Show Analyzer; the *total organic carbon can be calculated from the S_4 value.

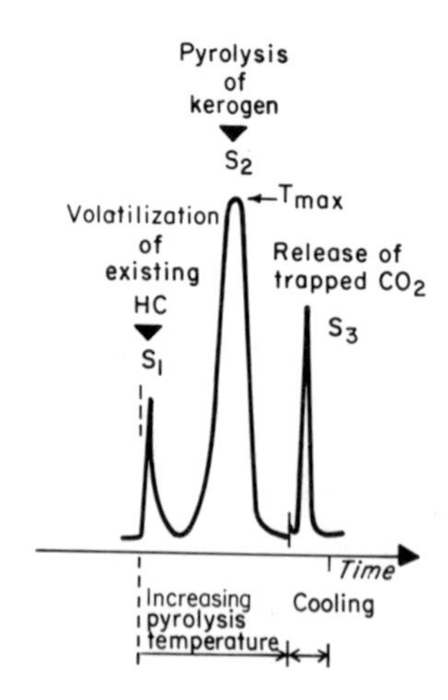

A typical Rock Eval output

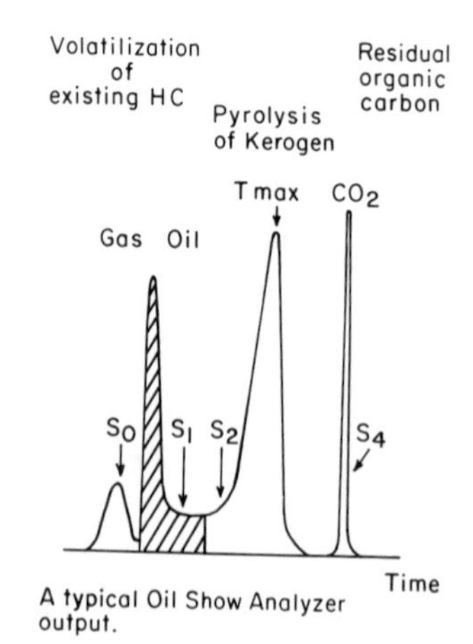

A typical Oil Show Analyzer output.

sapropel Amorphous, finely divided *organic matter derived from both plant and animal origin, which is deposited in sediments. It has been used in the context of *amorphous *oil prone kerogen in *organic geochemistry. There is minor use of the terms algal sapropel (to mean the chemical equivalent of *alginite or *Type I kerogen), and waxy sapropel (to mean the chemical equivalent of structured *liptinite or *Type II kerogen). *Cannel and *boghead coals are described as sapropelic coals. Synonymous with unstructured *liptinite, *amorphinite 1, and *amorphous kerogen. Equates chemically with Type I and Type II kerogen.

saturated hydrocarbons *Hydrocarbons which contain only carbon–carbon single bonds. Synonymous with alkanes.

saturate fraction The fraction of *crude oils and sediment *extracts which contains *normal, *branched, and *cyclic alkanes, eluted with pentane or heptane from *liquid chromatography. The relative proportion of the saturate fraction in oils and extracts increases with increasing maturity. The saturate fraction may be 20 to 30 per cent of a sediment extract or 30 to 80 per cent of a crude oil. Synonymous with paraffins and naphthenes and alkanes.

saturate fraction to total extract ratio The ratio of the total weight of the *saturate fraction to the weight of total extractable *hydrocarbons from a sediment. The value increases to about 20 to 40 per cent in oil prone sediments with increasing maturity. Values greater than 50 per cent are usually the result of contamination.

saturate to aromatic ratio The ratio of the *saturate fraction to *aromatic fraction weights in *crude oils as determined by *liquid chromatography. It is used primarily as a measure of maturity, but it is also a function of the *kerogen types in the source rock. The ratio may be changed by *water-washing and *biodegradation.

SCI Spore Colour Index.

sclerotinite The *coal petrographic term for one of the *inertinite group of *macerals which is derived from the fossilized remains of fungal sclerotia, hyphae, mycellia, and spores. It is opaque in transmitted white light, and grey to white and highly reflective in reflected white light. It can be found in tubular, cellular, or complex forms, and is commonly found in Tertiary brown *coals. It has no hydrocarbon potential at any level of maturity and equates chemically to *Type IV (IIIB) kerogen.

SCOT chromatography Support Coated Open Tube chromatography. A form of capillary column *gas chromatography.

secondary maceral A *maceral formed from *kerogen as a result of *maturation, e.g. *exsudatinite and *micrinite.

secondary migration The *migration of *hydrocarbons from the *source rock through permeable carrier horizons, along faults and unconformity surfaces (*conduits) to a trap (see Fig.). The major driving forces for secondary migration are capillary pressure, buoyancy, and pressure differential. Secondary *migration routes may be mapped using subsurface structure maps and *potentiometric maps. Chemical and physico-chemical processes acting on the hydrocarbons during secondary migration can cause compositional changes. These may occur as a result of solution of the more water-soluble, *aromatic compounds, and the *polar compounds in oils will be more readily absorbed on mineral surfaces. The relative abundance of *sterane *stereoisomers reputedly changes with secondary migration. Relative diffusion rates will cause compositional changes within gases; they become drier and *nitrogen-enriched with increasing distance from source. Diffusion, selective solution, and temperature- and pressure-dependent phase changes may cause carbon isotopic changes. *Reference*: Silverman, S.R. (1965). Migration and segregation of oil and gas. In *Fluids in subsurface environments—a symposium.* Am. Assoc. Petr. Geol., Mem. 4. Tulsa, Oklahoma.

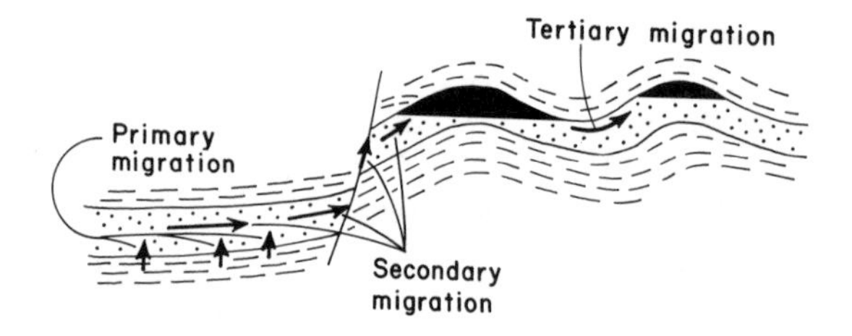

semifusinite One of the *inertinite group of *macerals, derived from partial oxidation of woody tissue. It is less oxidized than *fusinite and hence may have minor gas-generating potential at appropriate maturity levels. It is dark brown to opaque in transmitted white light, and grey to white in reflected white light. It has a lower reflectance than fusinite and higher relief; some original structure may be preserved. It equates chemically to *Type IV (IIIB) kerogen.

sesquiterpenoids Compounds made up of three *isoprene units, C_{15} compounds, which may be either cyclic or acyclic. They are derived from higher plants, resins, essential oils and some *bacteria. Sesquiterpanes, $C_{15}H_{24}$ compounds (See Fig., p. 113), form a significant part of terpenoids in the sedimentary record, especially products of the thermal decomposition of resinite. They are analysed by *gas chromatography–mass spectrometry of the *saturate fraction of oils and sediment *extracts using *m/e* 204. *Reference*: Simoneit, B.R. T., Grimalt, J.O., Wang, T.G., Cox, R.E., Hatcher, P.G., and Nissenbaum, A. (1986).

Cyclic terpenoids of contemporary resinous plant detritus and of fossil woods, ambers and coals. *Org. Geochem.*, **10**, 877–89.

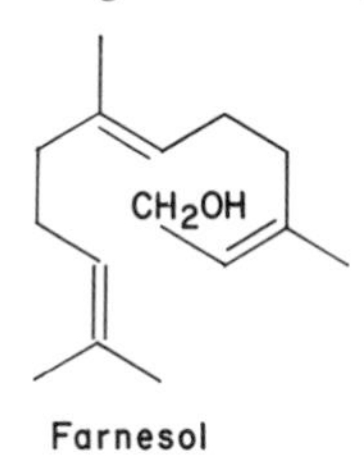

Farnesol

SIM Single Ion Monitoring. The use of *gas chromatography–mass spectrometry to monitor for a single fragment (*m/e*), to examine the occurrence of a specific class of compounds.

Single Ion Monitoring *See* SIM.

sitostane C_{29} *sterane. Synonymous with ethyl cholestane.

SMIM Selective Metastable Ion Monitoring. The double focusing (*GC–MS–MS) equivalent of *SIM or *MID. *See also*:metastable ions.

SMOW Standard Mean Ocean Water. The standard against which most *oxygen and *hydrogen *isotope abundances are most commonly measured. *See also*: NBS-18, NBS-19, PDB. (See Fig.).

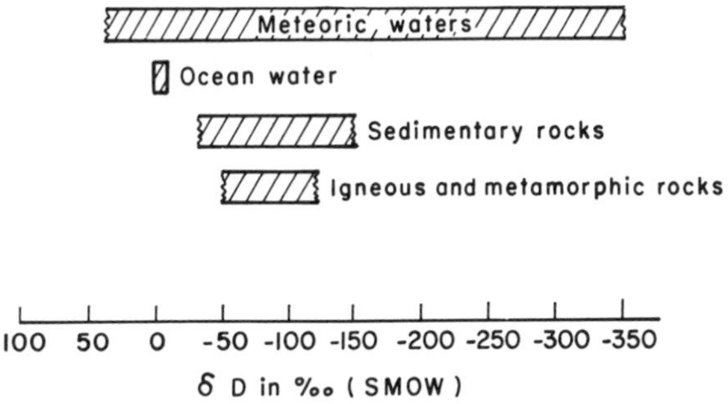

Meteoric waters
Ocean water
Sedimentary rocks
Igneous and metamorphic rocks

100 50 0 -50 -100 -150 -200 -250 -300 -350

δ D in ‰ (SMOW)

solidified bitumen A solid derived from oil or *bitumen by precipitation, de-asphalting, or *maturation of *kerogen. It may occur in *source rocks or reservoirs, or as veins, in pressure solution planes, *tar mats or as stylolites. It is largely insoluble in carbon disulphide (see Fig., p. 114). The reflectance of bitumen increases in response to increasing temperature. It reflects lower than *vitrinite up to a reflectance of ~1.1 per cent and then higher. It occurs in carbonate source rocks where it may be no more than a mobile form of *kerogen, there being no clays for the adsorption of the *organic matter. Synonymous with asphalt, grahamite, gilsonite, ozocerite, impsonite, ingramite, albertite and *pyrobitumen. *Reference*: Curiale, J.A. (1986).

Origin of solid bitumens, with emphasis on biological marker results. *Org. Geochem.*, **10**, 559–80.

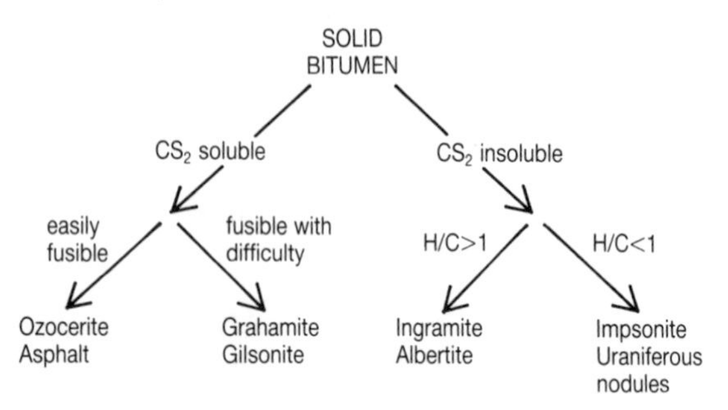

After Hunt *et al.* (1954). Origin of hydrocarbons in the Uinta Basin, Utah. *Bull. of Am. Assoc. Petr. Geol.*, **38** pp.1671–98.

SOM Sedimentary Organic Matter. *See* organic matter.

soluble organic matter *See* extract

source rock A rock which contains sufficient *organic matter of suitable chemical composition to generate and expel *hydrocarbons at appropriate maturity levels. *See also*: effective source rock, potential source rock.

soxhlet extraction A method for removing the soluble *organic matter from rocks using hot solvent refluxing (see Fig.). Finely crushed rock is held in a thimble usually made of cellulose or porous pot. The solvent,

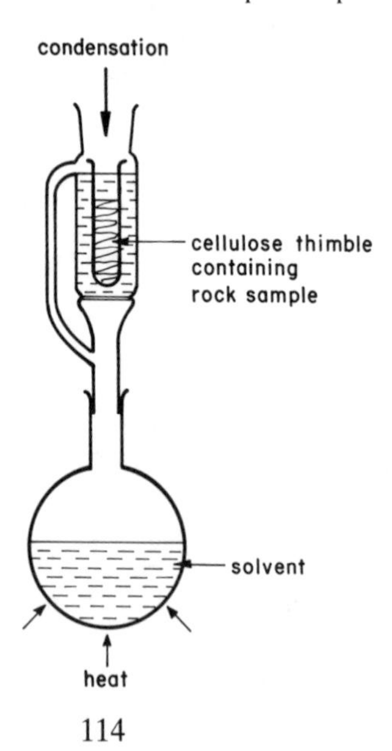

usually dichloromethane, is heated and the vapour is made to condense into a chamber containing the thimble and rock sample. An overflow system returns extract and solvent back to the solvent reservoir. The solvent is recycled and eventually all the extractable material is transferred to the solvent reservoir. Samples are usually refluxed for 24 hours. This used to be the standard method for extraction. *Ultrasonic extraction is used increasingly as it is quicker, but it is more labour-intensive.

Spore Colour Index, (SCI) A maturity measurement based on the colour of spores in *kerogen concentrates. It is a refinement of the Staplin *thermal alteration index (TAI) scale, and expands that part of the TAI scale which covers the *oil window.

Unornamented spores are used for colour measurement as they have uniform exine thickness and hence consistent colour. Many different scales exist, but the most frequently used is a scale of 1 to 10, which was calibrated against the original Staplin kerogen preparations. Using a system of yellow, red, and blue filters, the colour variations in spores, between totally *immature kerogen (yellow-green) to *post mature kerogen (black)were divided into ten scale divisions. A reference set of spores corresponding to these divisions was then used. It is possible to distinguish four grades of colour between these major scale divisions, so that there are a possible 32 divisions of spore colour. Readings are quoted as the mean of ~20 determinations on each sample. *Immature for oil generation on this scale, is equivlent to a spore colour of <3.5, *early mature 3.5 to 5.0, *peak mature 5.0 to 7.0, *late mature 7.0 to 9.0, and *post mature >9. Consistency is achieved by the distribution of sets of spore colour slides to different laboratories. Another less widely used scale is 1 to 7, with the onset of oil generation being equivalent to 3. Spore colour determinations, if performed competently, have several advantages. There is an *in situ* age check on the organic matter measured, as many spores are age-diagnostic. Spores produce *light oils at the appropriate maturity levels, so differences in *chemical kinetics for the oil generation process and reactions which produce colour changes in the kerogen are slight. Disadvantages are that spores can become darkened by trace metals and oil staining, or bleached by oxidation. *Reference*: Barnard, P.C., Cooper, B.A. and Fisher, M.J. (1976). Organic maturation and hydrocarbon generation in the Mesozoic sediments of the Sverdrup Basin, Arctic Canada. *4th Int. Pal. Conf. Lucknow*, 581–8.

sporinite The *coal petrological term for one of the *exinite group of *macerals, which is derived from the skins, or exines, of spores or pollen. It is yellow to brown in transmitted white light, and fluoresces in UV light. Initially it is less reflecting than *vitrinite, but its characteristics become similar to those of vitrinite when highly *mature. The most abundant constituents of sporinite are miospores, which are very small. The largest spores are the megaspores, which are dominantly of Carboniferous age. Chemically, sporinite is composed of condensed *aromatic rings but with considerable amounts of *aliphatic compounds; it produces *light oils at the appropriate maturity levels and equates chemically to *Type II kerogen.

sporopollenin A resistant, refractory organic biopolymer which forms the exines of spores, pollen, and dinocysts. It is a high molecular weight polymer of *carbon, *hydrogen, and *oxygen which is extremely physically durable and insoluble in acids, but is susceptible to oxidation. When preserved in sediments it produces gas and light oil at the appropriate maturity levels. It equates chemically to *Type II kerogen. *Reference*: Brooks, J. and Shaw, G. (1972). The chemistry of sporopollenin. *Chem. Geol.*, **10**, 69–87.

stable carbon isotopes Carbon has two stable isotopes, ^{12}C and ^{13}C. The measure of the relative abundance of these two isotopes is known as the stable carbon isotope ratio. The ratio is measured on *organic matter by converting all the organic matter to carbon dioxide, then measuring the relative abundances of the two isotopes using a mass spectrometer. The stable carbon isotope ratio, $\delta^{13}C$, is calculated from the formula:

$$\delta^{13}C‰ = \frac{{}^{13}C/{}^{12}C \text{ sample} - {}^{13}C/{}^{12}C \text{ standard}}{{}^{13}C/{}^{12}C \text{ standard}} \times 1000$$

and is expressed in units per mil or ‰.

The most widely used standard in petroleum geochemistry is *PDB, followed by *NBS-22. The relationship between these two is given by the formula:

$$\delta^{13}C(\text{PDB}) = 0.9702 \times \delta^{13}C(\text{NBS-22}) - 29.8.$$

Changes in the relative abundance of the two isotopes, known as *fractionation, occur by many different mechanisms; for example photosynthesis in green plants and *chemical kinetic effects. *Methanogenic bacteria cause one of the largest fractionations, producing *methane which is up to 40 per mil lighter than *thermogenic methane. Land plants are isotopically about 10 per mil lighter than marine plants due to differences in metabolic pathways. Cold water (2°C) plants are about 10 per mil lighter than warm water (15°C) plants. Different chemical groups within the plant material also have different isotopic abundances due to fractionation; the *lipid fraction is always isotopically lighter than the *aromatic fraction (see Fig.). In gases, additional isotopic fractionation can occur by diffusion as well as cracking so that differences in *migration (as well as of source and maturation) may be detected. During *diagenesis of organic matter the carbon isotope ratio changes, but still reflects the basic origin of the organic material; it can

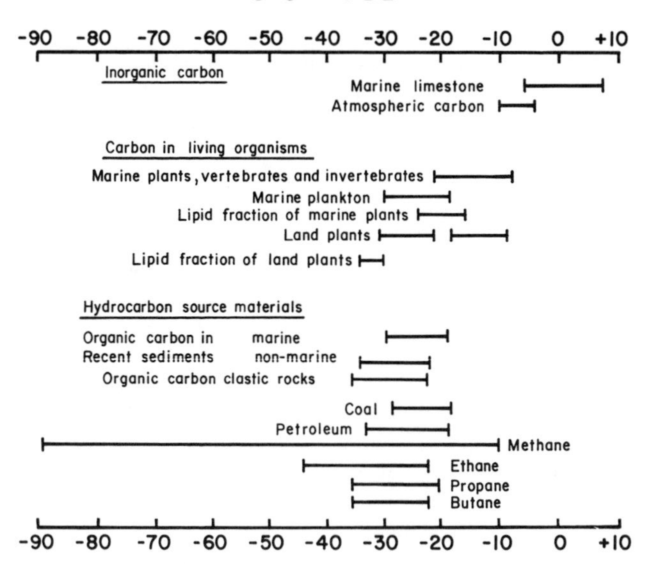

be used for kerogen typing and to determine the origin of soluble organic matter and oils. Oils are usually closer to the isotope ratio of the *lipid fraction of organic matter rather than the *total organic carbon isotope ratio. The absolute isotope ratio and the known changes which result from *maturation are used in *oil–source correlations. *See also*: secondary migration. *Reference*: Schoell, M. (1984). Recent advances in petroleum isotope geochemistry. *Org. Geochem.*, **6**, 645–63.

stem An informal and little used kerogen typing term which describes the *organic matter which is derived from plant stem. It shows distinct cellular structure, and equates chemically with *Type II to Type III kerogen.

steranes The alkanes derived from *steroid natural products. These compounds are found in the *saturate fraction of sediments and oils. *See also*: methyl steranes, rearranged steranes, regular steranes, steroids.

stereoisomers Compounds made up of the same atoms bonded by the same bonds but having different three dimensional structures, which are not interchangeable. These 3D structures are called configurations. *See also*: diastereoisomers, enantiomers, isomers. *References*: Organic chemistry textbooks.

steroids C_{27} to C_{35} three- or four-ring cyclic compounds made up of six *isoprene units, derived from plants and animals. They are detected by *gas chromatography–mass spectrometry of the *aliphatic and *aromatic fractions of oils and sediment extracts. *Steranes, which are found in oils and *source rock *extracts, are derived from sterols via sterenes and steradienes. Sterols are only found in biological material and shallow sediments, and steranes are not found in living *organic matter. The ratio of different steranes in oils and sediment extracts may indicate the type of organic matter which has sourced the steranes. C_{27} sterols are more abundant in *marine organic matter and C_{29} sterols are more abundant in higher land plants; C_{30} sterols may only occur in marine organic matter. The ratio of C_{27}:C_{28}:C_{29} steranes, commonly plotted on a triangular diagram, may indicate the *kerogen composition of source organic matter and is of use in *oil–source correlations (although this plot is not as diagnostic as was thought originally). They are highly optically active and have more than one *chiral centre, commonly at the C-20 and C-24 positions. Naturally-occurring sterols have $8\beta,9\alpha,10\beta,13\beta,14\alpha,17\alpha,20$-*R* stereochemistry. Thermodynamically induced changes to this basic stereochemistry occur at the 5,14,17 and 20 positions for *regular steranes, and 5,13,17, and 20 positions in *rearranged steranes, so that maturity ratios relate the concentrations of α and β configurations and the **R* and **S* isomers.

On increasing *maturation, the steranes rings may become *aromatic. A series of compounds, known as *aromatic steroids, are also useful maturation and oil–source correlation parameters. The two common types of aromatic steroids are the mono and tri aromatic steroids. These compounds are especially useful when oils have been biodegraded. Severe *biodegradation removes different classes of steranes at different rates. The biological configuration C-20 *R* compounds are removed faster than C-20 *S*. C_{27} steranes are removed faster than C_{28} and C_{29}. Aromatic steroids are the most resistant to biodegradation and provide a means of maturation and source correlation even when steranes have been completely removed. *See*, Table 1, summary section for maturation summary. *See also*: aromatic steroids, methyl steranes, rearranged steranes, regular steranes. *Reference*: Mackenzie, A.S., Brassell, S.C., Eglinton, G., and Maxwell, J.R. (1982). Chemical fossils, the geological fate of steroids. *Science*, **217**, 491–504.

straight chain alkanes *See* normal alkanes.

structural isomers Compounds with the same molecular formula but different structural formulae. Structural formulae ignore the 3D nature of a molecule and merely indicate that atoms are connected by a bond. Isomers may also result from differences in the position of a double bond or substituent group. They may be responsible for differences in reactivity. There is a geometric increase in the number of isomers as *carbon chain length increases, for example, *butane (C_4H_{10}) has two structural isomers whereas decane ($C_{10}H_{22}$) has 75. Melting and boiling points are usually lower for branched structures.

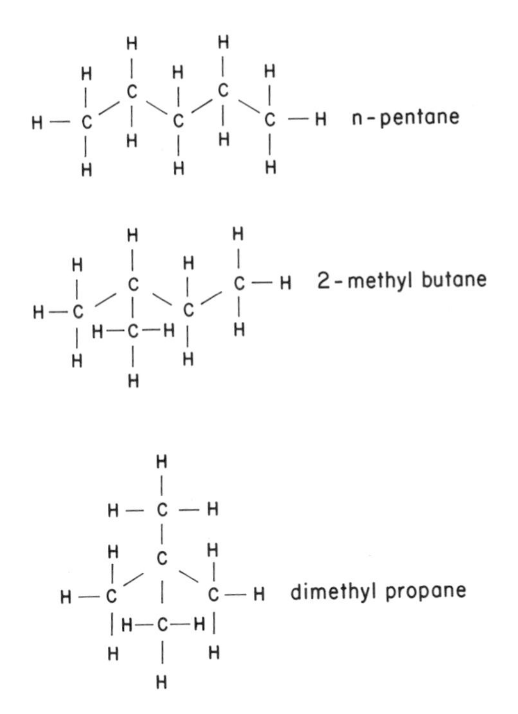

The isomers are named after the longest continuous carbon chain. Thus the three pentane isomers illustrated (see Fig.), are n-pentane, 2-methylbutane, and dimethylpropane. *References*: Organic chemistry textbooks.

structured liptinite *See* liptinite.

suberinite A *maceral of the *exinite group which is derived from cork cell walls of bark, and root stems of fruit (both as a protection against desiccation). It is derived from *cellulose, *lignin and suberin, and is similar to *cutinite. It shows weak reddish *fluorescence at low to medium maturity levels. It is rarely observed in fossil form, and equates chemically with *Type II kerogen.

suboxic *See* dysaerobic.

subsidence history The record of the downward settling of an horizon with minor or no horizontal movement, is called a subsidence history. Subsidence may be caused by a variety of mechanisms, such as faulting or thermal contraction. Observed subsidence may be the result of several different processes so that, to quantify subsidence, influences such as sediment loading and isostatic adjustments have to be considered. *Reference*: Roydon, L., Sclater, J., and von Herzen, R. (1980). Continental margin subsidence and heatflow—important parameters in the formation of petroleum hydrocarbons. *Am. Assoc. Pet. Geol. Bull.*, **64**, 173–87.

sulphate-reducing bacteria *Anaerobic bacteria which gain their energy for growth by coupling the oxidation of *organic matter to the reduction of sulphates (see Fig.) are known as sulphate-reducing bacteria; *hydrogen sulphide is a by-product of the reduction. The production of framboidal *pyrite is closely associated with the action of these bacteria, as the hydrogen sulphide which they produce reacts with iron to form trilolite and, eventually, pyrite. Sulphate-reducing bacteria are dominant over methanogenic bacteria until sulphates become exhausted. They are active in the uppermost layers of sediments; their abundance decreases rapidly with increasing depth. They are responsible for most of the destruction of organic matter in the anaerobic environment. *Reference*: Berner, R.A. (1985). Sulphate reduction, organic matter decomposition and pyrite formation. *Phil. Trans. R. Soc. Lond.*, **A 315**, 25–38.

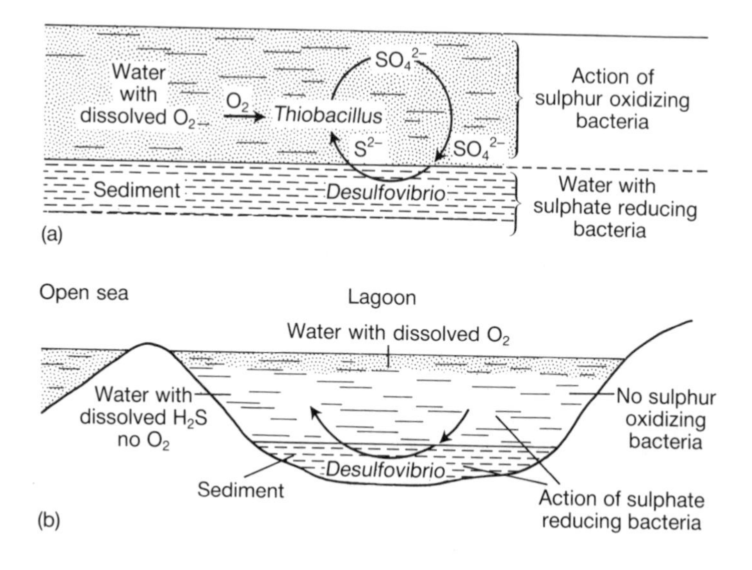

sulphur The element with atomic weight 32. It is a yellow solid at normal temperatures and pressures. It has five *isotopes ^{32}S, ^{33}S, ^{34}S, ^{35}S, and ^{36}S. It occurs naturally in the form of H_2S in sediments as a result of the action of *anaerobic bacteria. It occurs in oils and *natural gas accumulations as H_2S (oils and gases which contain significant amounts of H_2S are called 'sour'); additionally oils may contain sulphur compounds such as thiols, mercaptans, and thiophenes (see Fig., p. 121). Sulphur compounds occur in the *polar/NSO and *asphaltene fraction of oils. Most oils contain low amounts of sulphur (<1.0 per

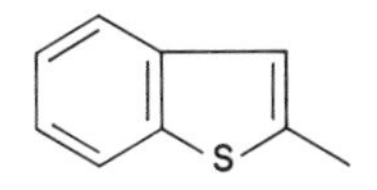

Methylbenzothiophene

cent), but some oils, e.g. from the Middle East, may have contents of 2 to 3 per cent. Sulphur contents of oils can be modified, but high sulphur is usually associated with *biodegradation, *water-washing, and derivation from carbonate source rocks. Carbonate-sourced oils tend to have higher sulphur content due to the lack of inorganic iron to complex with the sulphur.

sulphur isotopes There are four stable sulphur isotopes, ^{32}S, ^{33}S, ^{34}S, and ^{36}S, and one unstable isotope, ^{35}S. The major isotopic comparison used is between the relative abundance of ^{32}S and ^{34}S. The standard against which these abundances are measured is the *Canon Diablo meteorite triolite (see Fig.). The isotopic ratio of sulphur in sedimentary rocks varies from −40 to 50 per mil. *Fractionation of the sulphur isotopes occurs by both *kinetic and bacterial effects. H_2S produced from bacterial reduction of sulphate is isotopically *light.

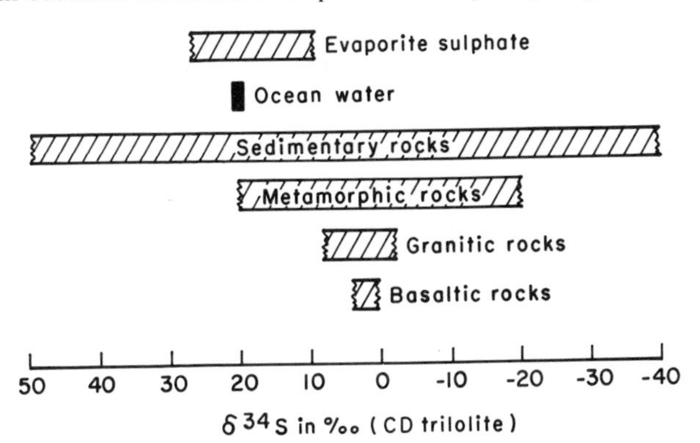

The main use of sulphur isotopes in *organic geochemistry is in determining the origin of sulphur in oils and *natural gases. They enable the discrimination of sulphur from *marine and *non-marine sources. *Maturation does not always change the isotope ratio in oils and *kerogen. *Reference*: Orr, W.L. (1974). Changes in sulphur content and isotope ratio of sulphur during petroleum maturation—a study of Big Horn Basin Palaeozoic oils. *Am. Assoc. Pet. Geol. Bull.*, **58**, 2295–318.

surface geochemistry The detection of *hydrocarbon accumulations at depth using surface manifestations, such as seeps of oil and gas. Soil or sediment gases can be examined for composition, and stable carbon isotope ratios measured for evidence of thermogenic genesis; aromatic compounds may also be used. In addition to searching for direct evidence of seeping gases, anomalous concentrations of *bacteria which metabolize the *light hydrocarbon gases can be detected, and their abundance mapped. *Reference*: Philp, R.P. and Crisp, P. T. (1982). Surface geochemical methods used for oil and gas prospecting—a review. *J. Geochem. Explor.*, **17**, 1–34.

TAI *See* Thermal Alteration Index.

tar *See* asphalt.

tar mat The concentration of *heavy oil at, or near, the oil–water contact in a reservoir is called a tar mat. Tar mats result from in-reservoir alteration of oils, by *maturation, *biodegradation, *water-washing, or gas-deasphalting. They comprise oil enriched in *asphaltenes, which can be generated *in situ*, or accumulate by gravity segregation. *See also*: asphaltenes. *Reference*: Dahl, B. and Speers, G.C. (1985). Oseberg Field tar mat. *Org. Geochem.*, **10**, 547–58.

tasmanites A species of alga. Its remains may form a recognizable part of *alginite. *See also*: fluorescence.

T_{eff} An abbreviation for effective heating time. *See also*: Hood *et al.*

telinite One of the *vitrinite group of *macerals identified by microscopic examination of *kerogen. Telinite is vitrinite which shows the distinct cellular structure of woody tissue, and the cells may be filled with *collinite or *resinite. It is reddish brown in transmitted white light, does not fluoresce in UV light, and is grey in reflected white light. Its reflectance increases with maturity as a result of increasing aromatization. Synonymous with telovitrinite. It equates chemically to *Type III kerogen. *See also*: collinite.

terpanes Abundant, biologically derived *alkanes made up of two *isoprene units, C_{10}; also known as monoterpanes. They are found in oils and sediments, and are derived from the *terpenoids of *bacteria and land plants. Loosely used to include *diterpanes and *triterpanes.

terpenoids The group term which describes *alkanes, *alkenes, alcohols etc., derived from two *isoprene units. Loosely used to describe all related classes of natural products and *biological markers, irrespective of number of carbon atoms.

terrestrial Sedimentary deposits laid down above tidal reach are called terrestrial. *Organic matter is said to be terrestrial if it is derived from land plants, but it may be included in a marine sediment. An oil may be of apparent terrestrial origin, e.g. waxy and of *light *carbon isotope ratio but from sedimentologically marine sediments, an especially common occurrence in deltaic environments. Markers of the terrestrial origin of oils and *extracts may be abundant *pristane and *phytane, abundant C_{25} to C_{35} *normal alkanes, usually with a marked *odd–even predominance, bicadinane, sesquiterpanes from resins, *18α(*H*)-oleanane, *diterpenoids from higher plants, abundant *steranes relative to *hopanes, and abundant C_{29} steranes relative to C_{27} and C_{28}. Two unidentified compounds, which are terpenoid hydrocarbons, called *X and *Y in the literature, are also thought to be indicators of terrestrial derivation.

tertiary migration The name given to the remigration of oil and gas from an existing site of accumulation (see Fig.). It may occur as a result of Gussow displacement by later-generated oil or gas into a spillover system, or by tectonic readjustment of an existing structure.

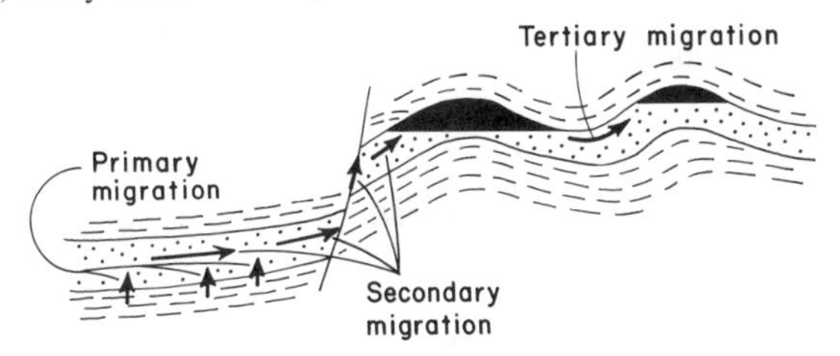

tetracyclic terpanes C_{24} to C_{27} four-ringed compounds found in oils and sediment *extracts. They may be bacterially derived, and appear to be more abundant in Palaeozoic aged samples. They occur in both *marine and *non-marine organic matter. The most common member of the series is the C_{24} compound. They are detected using *gas chromatography–mass spectrometry of the *saturate fraction with a *diagnostic ion of *m*/*e* 191.

2,6,10,14-tetramethylhexadecane *See* phytane.

2,6,10,14-tetramethylpentadecane *See* pristane.

theoretical maturity An estimate of maturity produced from thermal and *kinetic models of basin development and *kerogen transformation. *See also*: maturation modelling.

Thermal Alteration Index, (TAI) A maturity scale based on *kerogen colour, usually from spores and pollen, developed by Staplin. Kerogen concentrates are examined in transmitted white light, and their colour assessed and determined relative to a standard. The scale of 1 to 5 may have decimal subdivisions, 1.5, 1.75, or gradational subdivisions 1+, 2−, etc. The scale was developed for a wide range of maturity levels, most of which were in the *late mature to *post mature region. *Immature for oil generation equates to <2.2, *mature 2.2 to 3.5, and post mature >3.5. For *gas prone *source rocks immature is >2.5, mature 2.5 to 4.0, and post mature <4.0. As colour intensity varies with thickness of kerogen, determinations may not be as consistent as with spore colour. *See also*: Spore Colour Index. *Reference*: Staplin, F.L. (1969). Sedimentary organic matter, organic metamorphism and oil and gas occurrence. *Bull. Can. Pet. Geol.*, **17**, 47–66.

thermal conductivity (k) A measure of the ability of materials to conduct heat. Metals have high thermal conductivity; asbestos has a low thermal conductivity, and geological materials also show relatively low values. Chalk and salts have the highest thermal conductivities, shales have low thermal conductivities. Water decreases thermal conductivity, so that compacted rocks are more conductive than uncompacted rocks. Typical thermal conductivities range from 3 to 15×10^{-3} cal/cm s °C. Thermal conductivity and *heatflow determine *geothermal gradients. These properties are used in thermal models for *maturation modelling. *Reference*: Robertson, E.C. (1979). Thermal conductivity of rocks. *USGS open report*, 79–356.

thermal gradient *See* geothermal gradient.

thermogenic gas *Hydrocarbon gases which are sourced from *organic matter by thermal breakdown, usually at depth (see Fig.). These gases

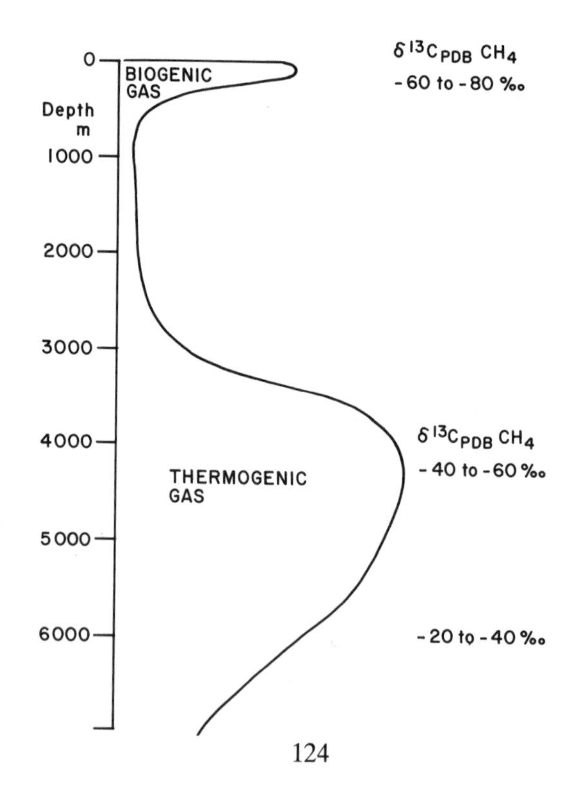

have stable carbon isotope ratios from −20 to −40 per mil. They are derived from *mature *gas prone or *post mature *oil prone kerogen. They are normally dry gases, although they may contain hydrocarbons in the gasoline range if they are from a post mature oil source. *Reference*: Schoell, M. (1983). Genetic characterisation of natural gases. *Am. Assoc. Pet. Geol. Bull.*, **67**, 2225–38.

Thin Layer Chromatography *See* TLC.

thiourea A sulphur-containing organic compound used for adduction in the preparation of *sterane and *triterpane concentrates for *gas chromatography–mass spectrometry, formula NH_2CSNH_2. It forms chains of hexagonal crystals which act like *molecular sieves. It adducts *isoprenoids, small ring compounds and some steranes, such as *cholestane.

Tissot and Espitalié Model A method for calculating the *theoretical maturity of *source rocks using *activation energy (E_a) distributions, applied via the *Arrhenius rate equation. The original method (see Fig.) used six activation energies ranging from 10 to 80 kcals/mol, whose distribution varied for each of the three hydrocarbon-generating chemical kerogen *Types I, *II, and *III. The activation energies were theoretical, and based on differences in chemical composition and structure. The range has been modified recently to a much narrower one of approximately 45 to 55 kcals/mol. Newer versions of the method use the P_2 peak from *Rock Eval measurements, to which a distribution of up to 18 activation energies is fitted using mathematical curve-fitting techniques. The maturity calculated is expressed as a percentage, or as a decimal value of 1.0, and is known as the *conversion index. A

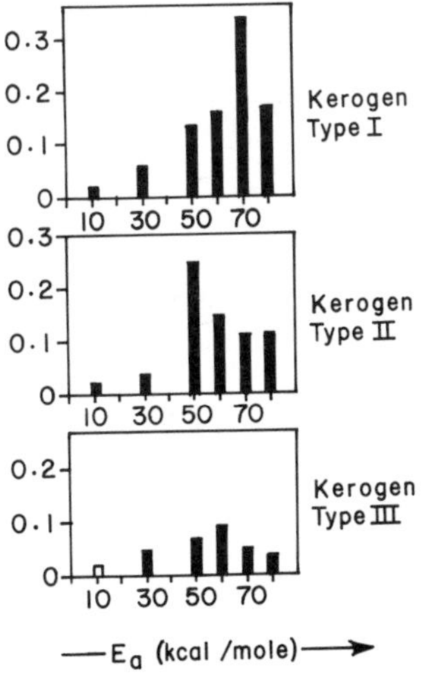

simplified approach is to use a single activation energy applied directly in the Arrhenius rate equation. The single activation energy can be measured directly from *pyrolysis experiments if suitable samples are available. *See also*: pseudoactivation energy. *Reference*: Tissot, B. and Welte, D. (1984). *Petroleum formation and occurrence* (2nd edn), 699 pp. Springer-Verlag, Berlin.

TLC Thin Layer Chromatography. A type of *liquid chromatography in which the stationary phase is deposited as a thin layer on a glass or plastic plate. A small amount of sample is added to the bottom of the plate, which is then suspended in an appropriate solvent in a closed container. The solvent diffuses up the plate through the sample, and various components move up the plate in accordance with partition coefficients. The technique is used for separating saturated from unsaturated compounds in the *aliphatic fraction of oils and *extracts, especially when only small quantities are available for analysis. *See also*: column chromatography.

T_m $C_{27}H_{46}$ *hopane, trisnorhopane, of 17α(*H*),22,29,30-tris-nor structure. It is used as a *maturation indicator in the ratio $T_m:T_s$, which can only be used within a series of samples from a consistent source, as the ratio may be strongly facies-dependent.

T_{max} The temperature, in °C, at which the pyrolytic yield of *hydrocarbons from a rock sample reaches its maximum, using *Rock Eval or *Oil Show Analyzer instruments. It is a *maturation parameter which is also *kerogen* type dependent. Approximate maturity boundaries for oil generation are <435 °C (*immature), 435 to 460 °C (*mature), and >460 °C (*post mature). If the S_2 is small, T_{max} is not measured reliably. The correlation of T_{max} with *vitrinite reflectance is shown in the Figure below. *See*, Table 1, summary section for maturation summary.

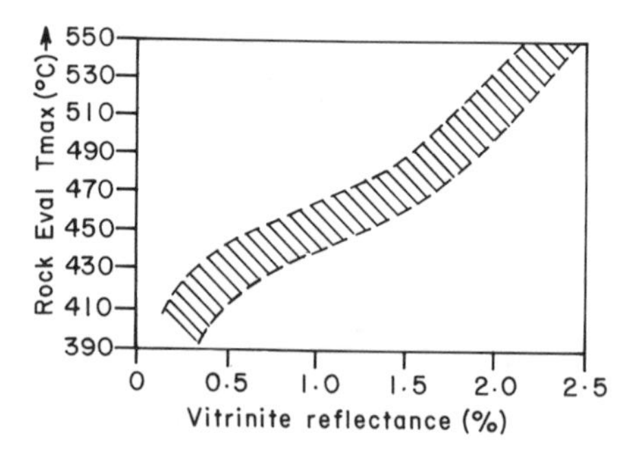

TOC *See* Total Organic Carbon.

toluene An organic compound which is liquid at room temperature, formula $C_6H_5CH_3$ (see Fig.). It has similar properties to *benzene but is

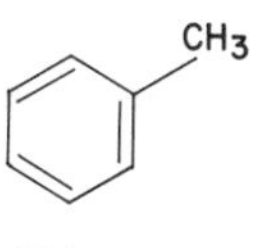

Toluene

less carcinogenic, so is used in place of it in many geochemical analyses. It is often used as the solvent for *aromatic compounds in *liquid chromatography. It is present in most oils; its absence is usually indicative of *water-washing. Synonymous with methylbenzene.

torbanites Coals which are composed of algal remains, of the type found in the Carboniferous of the Midland Valley of Scotland, and the Permian of Australia. They may generate waxy oils when *mature, but are usually only minor sources owing to their limited distribution. Coorongite is presumed to be a present-day equivalent. *See also*: alginite, boghead coals.

total extract/organic carbon ratio *See* extract to organic carbon ratio.

Total Organic Carbon, (TOC) A measure of the organic carbon in a rock, expressed as weight per cent, used as a fundamental parameter in classifying source rocks in conjunction with *kerogen type and *maturation. It is normally determined by heating samples (from which inorganic carbonates have been removed by acid digestion) to 1000 °C in an oxygen supply, and collecting and weighing the carbon dioxide thus created. Only small amounts of sample are required and the method is inexpensive and accurate. Any inaccuracies are caused when dolomite, which is more difficult to remove than calcite, remains in the sample, or if there are appreciable amounts of *sulphur present. As only a small sample is analysed, care has to be taken to ensure that the sample is representative. Guidelines for interpretation of values, providing the *kerogen will produce *hydrocarbons are:

Average all shales	0.9%
Average source shales	2.2%
Average source calcareous shale	1.9%
Average carbonate source	0.7%
Average all source rocks	1.8%

*Coals and oil shales have very high TOCs, usually 10 to 50 per cent. A problem with high TOC *oil prone sediments is that hydrocarbons may be absorbed by the *organic matter, and thus unable to migrate out of the source rock. Synonymous with Corg. *Reference*: Tissot, B. and Welte, D. (1984). *Petroleum formation and occurrence* (2nd edn), 699 pp. Springer-Verlag, Berlin.

***trans–cis* isomerism** *See* geometrical isomerism.

transformation ratio A ratio related to the extract to total organic carbon ratio and production index. It is rarely determined in practice but has significance in modelling of volumes of *hydrocarbons from source rocks. It is calculated from the formula:

$$\text{transformation ratio} = \frac{X_0 - X}{X_0}$$

where X is the hydrocarbon-generating potential of a sample, and X_0 is its original potential prior to *maturation. Units are mg hydrocarbon/g organic carbon; values vary from 0 to 1. *Reference*: Yukler, M.A. and Kokesh, F. (1985). A review of models used in petroleum resource estimation and organic geochemistry. In *Advances in Organic Geochemistry I* (ed. J. Brooks and D. Welte), 344 pp. Academic Press, London.

tricyclic terpanes Three-ringed compounds from C_{19} up to C_{45}. They appear to be derived from bacterial precursors, much as *hopanes. They have an asymmetric carbon from the C_{26} member upwards and have **R* and **S* *isomers. They are present in both *marine and *non-marine sediments and oils. Their relative abundance is enhanced by increasing maturity. They are particularly abundant in Palaeozoic-sourced oils. They are attacked by *bacteria in the same way as hopanes, and at the same severity of *biodegradation of oils. They are detected using *gas chromatography–mass spectrometry and have a *diagnostic ion of 191. Demethylated tricyclic terpanes have a diagnostic ion of 177. *Reference*: Aqino Neto, F.R., Trendle, J.M., Restle, A., Connan, J., and Albrecht, P. (1981). Occurrence and formation of tricyclic and tetracyclic terpanes in sediments and petroleums. In *Advances in Geochemistry* (ed. M. Bjorøy *et al.*) pp. 659–67. John Wiley, Chichester.

triterpanes C_{27} to C_{35} five ring cyclic *alkanes made up from six *isoprene units and derived from triterpenoid hydrocarbons in *bacteria, fungi, lichen, *algae, and higher plants (see Fig.). They are detected by *gas chromatography–mass spectrometry of the *saturate fraction of oil and sediment extracts, most with a *diagnostic ion of 191. They are *biological markers and highly optically active at the C-22 position in the C_{31} compound and higher homologues. Triterpanes are found in *organic matter, oils and sediment extracts. The biological stereo-

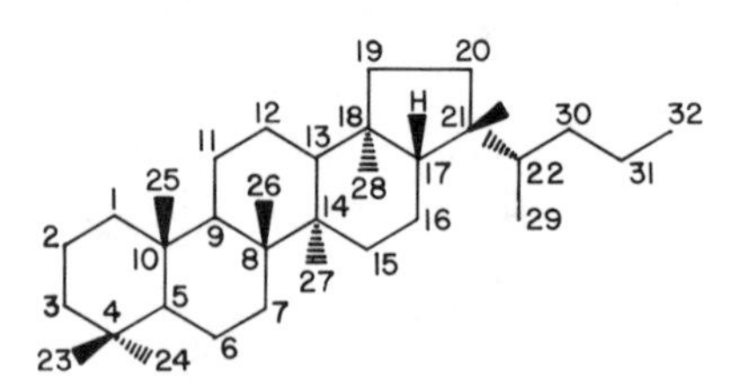

Hopane / Moretane

chemistry is 17β, 21β, 22-*R*. There is a thermodynamically controlled change to 17α, 21β, 22-*R* and 22-*S*, and 17β, 21α, 22-*R* and 22-*S*, so that the relative abundances of these compounds are used as *maturation parameters.

Within the triterpane family there are several groups of compounds of different origins. The dominant triterpanes in sediments and oils are the *hopanes, derived predominantly from bacteria; isomers of these compounds are the *moretanes. Other notable compound types are the *oleananes, 18α(*H*) being found only in Cretaceous and younger sediments, believed to be derived from angiosperms; gammacerane, whose abundance appears to be controlled by the salinity in *lacustrine settings; and lupanes, ursanes, and arboranes, derived from higher land plants. The ratio of 17α(*H*)-hopane (C_{30}), to C_{27} to C_{29} steranes is used as an indicator of bacterial organic matter. The ratio of T_m, trisnorhopane, to T_s, trisneonorhopane, is also used as a maturation parameter within samples from a single source. *See also*: hopanes, moretanes.

T_s Trisnorneohopane, formula $C_{27}H_{46}$, 18α(*H*),22,29,30-tri-nor structure. It is a source-dependent parameter in the ratio T_s:T_m, which is used as a *maturation ratio.

TSE Total Soluble Extract. *See* extract.

TTI Time Temperature Index. A term used in Waples' adaptation of the *Lopatin method for calculating *theoretical maturity in *maturation modelling. The index is a sum of values calculated for each 10°C segment of the temperature history of a source rock. The time spent in each segment is calculated from *burial history and *geothermal gradient data. This is then multiplied by a factor of 2, which is dependent on the exact temperature segment e.g. 100 to 110°C is 2^0, 150 to 160°C is 2^5, etc. The TTI should be calibrated against maturity data but, in the absence of data, default values are: 15 for onset of oil generation, 75 for peak oil generation, and 160 for the end of oil generation. *See also*: Lopatin.

turbodrilling A type of rotary drilling in which the drill bit is driven by a down hole turbine powered by mud flow. This type of drilling produces faster r.p.m. speeds, and hence higher temperature in the rock formations being drilled. Early use of this technique produced problems in geochemical samples. The high temperatures artificially matured the *kerogen, causing erroneous maturity results, and sometimes reducing *pyrolysis yields. When oil-based drilling muds were used in conjunction with turbodrilling *alkenes were notable mud-decomposition products. Turbodrilling techniques have now changed and these problems are rarely encountered. *Reference*: Taylor, J.C.M. (1983). Bit metamorphism, illustrated by lithological data from the German North Sea wells. *Geol. en Mijnbouw*, **62**, 211–19.

Type I kerogen A *kerogen type based purely on the chemical composition of the kerogen, i.e. on C, H, and O content (derived from *elemental analysis or *pyrolysis, especially *Rock Eval). The kerogen is hydrogen-rich and of an *aliphatic nature with *normal alkanes in greater abundance than *cyclic alkanes. *H/C ratios and *hydrogen indices of *immature Type I kerogen are >1.5, and 600 to 950 respectively; the *O/C ratio is <0.1.

It is derived from algal material preserved in *anaerobic environments, especially *lacustrine. It produces oil at the appropriate maturity levels. It is the chemical equivalent of *alginite, *amorphous kerogen, *amorphinite l, and *sapropel. The type example of this kerogen is from the non-marine, Eocene, Green River Formation, Utah, USA (see Fig.). *Reference*: Katz, B.J. (1983). Limitations of Rock Eval pyrolysis for typing organic matter. *Org. Geochem.*, **4**, 195–9.

Type II kerogen A *kerogen type based purely on the chemical composition of the kerogen, i.e. on C, H, and O content (derived from *elemental analysis or *pyrolysis, especially *Rock Eval). The kerogen is of comparatively hydrogen-rich, alicyclic or naphthenic nature, typically derived from exinitic debris or degraded phytoplankton, which were preserved in *anaerobic to *dysaerobic environments. It contains higher amounts of *sulphur than the other kerogen types. *H/C ratios or *hydrogen indices of *immature Type II kerogen are 1.5, and 400 to 600 respectively. It produces oils at appropriate maturity levels although it is of lower potential than *Type I kerogen.

Type II kerogen is the chemical equivalent of *amorphous, *amorphinite l, *sapropel, *cutinite, *sporinite, *resinite, *exinite, *liptinite,

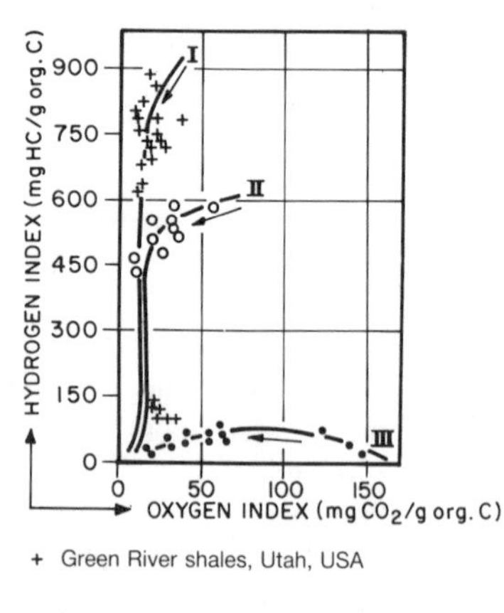

+ Green River shales, Utah, USA

o Lower Toarcian shales, Paris Basin, France

• Upper Cretaceous shales, Douala Basin, Cameroon

*suberinite and *herbaceous kerogen. The type example of this kerogen is from the marine Lower Toarcian Shales of the Paris Basin, France (see Fig., p. 130). *Reference*: Katz, B.J. (1983). Limitations of Rock Eval pyrolysis for typing organic matter. *Org. Geochem.*, **4**, 195–9.

Type III kerogen A *kerogen type based purely on the chemical composition of the kerogen, i.e. on C, H, and O content (derived from *elemental analysis or *pyrolysis, especially *Rock Eval). The kerogen is relatively hydrogen-poor, has a polyaromatic nature and is derived predominantly from higher plants. *H/C ratios and *hydrogen indices of *immature Type III kerogen are relatively low, i.e. <1.0 and 300 respectively. It produces gas and sometimes associated *condensate at appropriate maturity levels.

Type III kerogen is the chemical equivalent of *vitrinite, *telinite, *collinite, *huminite, *amorphinite v, *woody, *stem, and *humic kerogen. The type example is from the Upper Cretaceous shales of the Douala Basin, Cameroon (see Fig., p. 130). Sometimes called Type IIIA kerogen. *Reference*: Katz, B.J. (1983). Limitations of Rock Eval pyrolysis for typing organic matter. *Org. Geochem.*, **4**, 195–9.

Type IV kerogen An informal term for residual or oxidized kerogen based on its chemical composition, i.e. on C, H, and O content (derived from *elemental analysis or *pyrolysis, especially *Rock Eval). The kerogen is oxidized and hydrogen-poor, has an *aromatic, graphitic structure with a total absence of *aliphatic compounds. *H/C ratios and *hydrogen indices of *immature or mature Type IV kerogen are 0.5 and <50 respectively, and the O/C ratio is 0.2 to 0.3.

Type IV kerogen is the chemical equivalent of *inertinite, *fusinite, *semifusinite, *sclerotinite, *coaly, and *dead carbon. It has no *hydrogen indices of *immature or *mature Type IV kerogen are 0.5 and and <50 respectively, and the O/C ratio is 0.2 to 0.3.

UCM Unresolved Complex Mixture. Part of the *saturate or *aromatic *fraction of soluble *organic matter which is not resolved into peaks during *gas chromatography. Informally known as the 'hump' (see Fig., p. 132). In the saturate fraction these compounds are usually iso (*branched) and *cyclic alkanes.

ultrasonic extraction A technique used to obtain the soluble organic matter from sediments. The rock is crushed and mixed with an *organic solvent, such as dichloromethane. The mixture is immersed in an ultrasonic bath for a period of time ranging from several minutes to an hour. The solvent may have to be renewed several times for complete

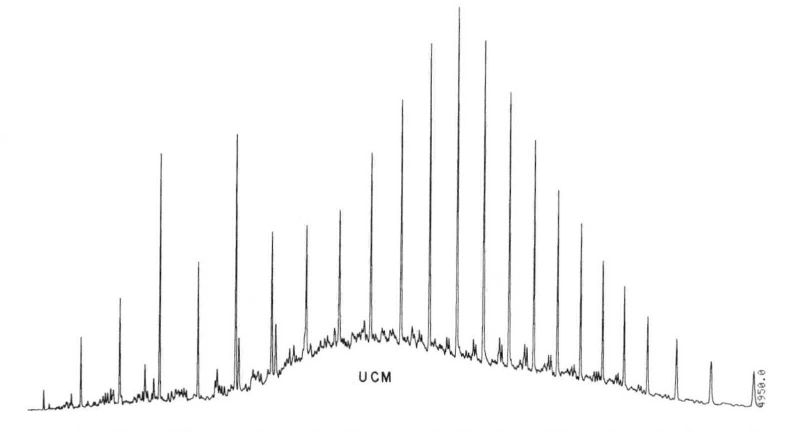

extraction. The rock and solvent mix is then filtered and the rock extract obtained by evaporation of the solvent. This method is quicker than hot solvent refluxing. *See also*: soxhlet extraction.

unstructured liptinite The *maceral group describing *oil prone *amorphous kerogen. It fluoresces in UV light. Synonymous with *amorphinite I and *liptodetrinite. It equates chemically with *Type II kerogen.

urea A solid white, crystalline organic compound, formula NH_2CONH_2. It is used in *organic geochemistry to adduct *iso- and *normal- *alkanes and alkyl benzenes from the *saturate fraction of *crude oils and *extracts. This enables the concentration of larger molecules, such as *steranes. The urea forms chains of hexagonal crystals which act in the same way as *molecular sieves. *See also*: thiourea.

ursanes *See* triterpanes.

Van Krevelen diagram The diagram used to display *O/C and *H/C ratios in *elemental analysis of *organic matter is called a Van Krevelen diagram (see Fig., p. 133). *Rock Eval *hydrogen and *oxygen indices are displayed on a plot which simulates the Van Krevelen diagram. *Reference*: Van Krevelen, D.W. (1984). Organic geochemistry old and new. *Org. Geochem.*, **6**, 1–10.

vitrinite The *maceral group derived from the lignified tissues of higher land plants, i.e. trunks, branches, stems, leaves, and roots of trees and plants. It is composed chemically of *aromatic nuclei with various

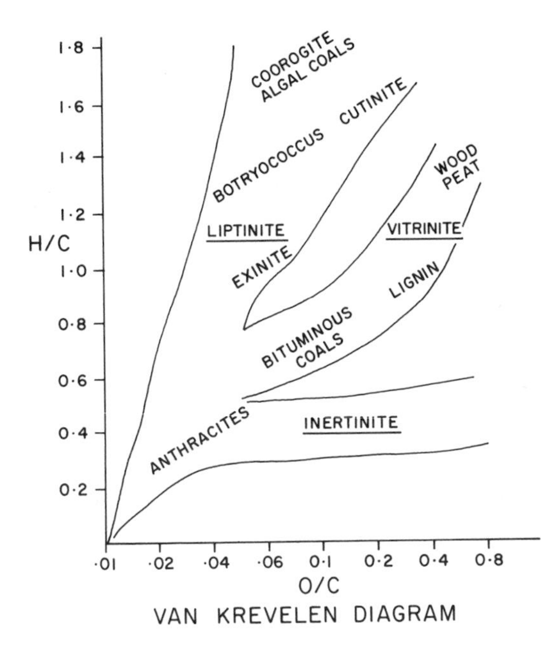

VAN KREVELEN DIAGRAM

functional groups. Vitrinite is derived from the *humic acid fraction of *humic substances, from predominantly *lignin and *cellulose. The environment of preservation is usually weakly acidic and reducing. It is light orange to dark brown in transmitted white light, and grey to yellow in reflected white light. It does not fluoresce in UV light.

Three types of vitrinite are defined, unstructured vitrinite or *collinite, structured vitrinite or *telinite, and detrital vitrinite or *detrovitrinite. *Bacteria may alter these forms to give a structureless or *degraded vitrinite. The change in the reflectivity of vitrinite with increasing temperature, caused by increasing aromatization, is used for *maturation measurement. Synonymous with woody, and partly humic. It produces gas at optimum maturity and is equivalent chemically to *Type III kerogen. *See also*: collinite, degraded vitrinite, telinite, vitrinite reflectance.

vitrinite reflectance A *maturation parameter based on the change in the reflectance of polished *vitrinite particles with increasing time and temperature; common abbreviations for vitrinite reflectance are VR_o, R_o, R_m, R_{max}, R_{min}, R_{mean}, $\%R_o$. Increases in the reflectance are caused by the progressive aromatization of the *kerogen with accompanying loss of *hydrogen in the form of *hydrocarbon gases; the end product of the process is graphite. Reflectance follows a logarithmic increase with depth in areas of constant *heatflow. This enables extrapolation of reflectance trends beyond sampled depths; hence, it may be used predictively. Although vitrinite itself can only produce gas, the *kinetics of reflectance changes are sufficiently like those of oil production from *oil prone kerogen to allow its use for prediction of oil generation, in most burial/temperature regimes.

Whole rock samples or kerogen concentrates are mounted in epoxy resin and polished using various grades of alumina and jewellers rouge. A microscope, (fitted with a photomultiplier and a data handling system), is used to locate the vitrinite particles in the sample. Their reflectance, usually under oil immersion (R_o), is then measured, and expressed as a percentage of incident light. Standards are used to calibrate the whole microscope assembly. Between 20 and 100 readings are considered necessary on each sample, depending on whether it is a whole rock sample or kerogen concentrate. These values are plotted as a histogram (see Fig.). The reflectance values are spread in Gaussian distributions or populations, from which a mean and standard deviation are calculated. The spread within a population is mostly due to particle orientation; this anisotropy becomes especially pronounced at reflectances greater than 1 per cent. Several populations may be present in one sample owing to *reworked or *allochthonous vitrinite and, in drill cuttings, to *lignite from drilling mud and down-hole caving.

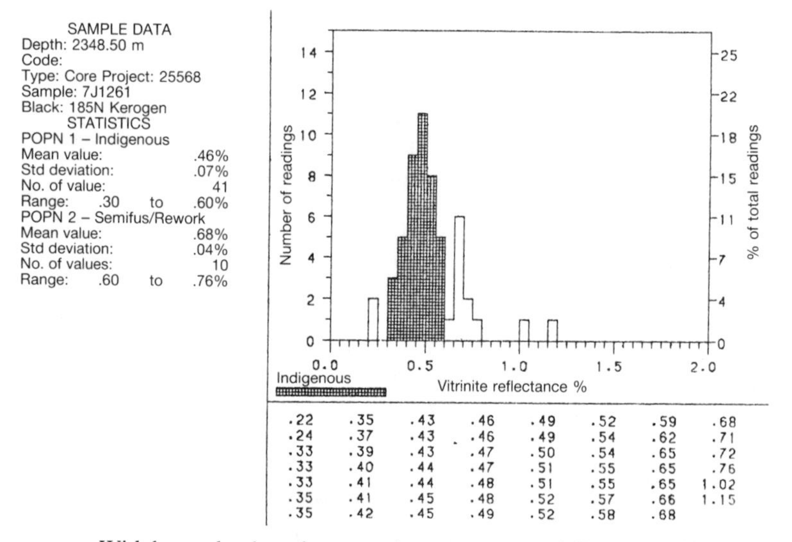

Widely used values for maturity zones are: <0.5 per cent *immature for oil generation; 0.5 to 1.3 per cent *mature for oil generation, and >1.3 per cent *post mature for oil generation. For gas, the equivalent maturity zones are <0.7 per cent immature, 0.7 to 3.0 per cent mature; and >3.0 per cent post mature.

In rapidly subsiding basins, discrepancies have been noted between oil production and predicted maturity from vitrinite reflectance. This is most probably due to the slight kinetic differences between oil prone kerogen transformation and reflectance increase, which is only apparent in extremes of temperature/time conditions. Errors in reflectance determinations occur if vitrinite is misidentified, or if the reworked or allochthonous material is confused with *in situ* or *autochthonous material. *See*, Table 1, summary section, for maturation values. *Reference*: Bostick, N.H. (1979). Microscopic measurements of catagenesis of solid organic matter in sedimentary rocks to aid exploration for petroleum to determine former burial temperatures—a review. *Soc. Econ. Pal. Min.*, Special publication No. **26**, 17–43.

vitrodetrinite Vitrinite which occurs in the form of irregular-shaped detritus from plant fragments which are degraded at an early stage.

volatile matter A *coal rank *maturation parameter, complementary to the *FCC (Fixed Carbon Content) scale. It may be found as a *source rock maturation parameter in older geochemistry reports; 40 per cent volatile matter is 60 per cent FCC, equivalent to a *vitrinite reflectance of 0.7 per cent, and 25 per cent volatile matter is 75 per cent FCC, equivalent to a vitrinite reflectance of 1.3 per cent.

water table maps *See* potentiometric maps.

water-washing A process whereby formation water, or ground water selectively remove the more soluble components of oil, i.e. lighter *hydrocarbons and *aromatic compounds and cause a lowering of *API gravity. The absence of *benzene and *toluene in oils is considered to be diagnostic of water-washing. Compaction water from *source rocks and formation water along migration routes is saturated with the water-soluble hydrocarbons, so when an oil is water-washed it indicates that it has been in association with water from a different source. Water-washing may occur during *migration, in a trap, during remobilization of an oil due to tectonic movements, during overspill, gas flushing (*tertiary migration) or during hydrodynamic flow. It is commonly, although not exclusively, associated with *biodegradation, as the new water may be fresh and carrying active *bacteria. *See also*: biodegradation.

waxes Part of the *lipid fraction of *organic matter. They are usually from the protective coating of leaves, and comprise long chain *normal alkanes and alcohols, with chain lengths from C_{16} to C_{36}; there is a predominance of *alkanes in the C_{25} to C_{35} region. In sediments, alcohols undergo alteration to normal alkanes. The term is used informally to describe alkanes which are solid at room temperature, and thus cause high pour points in *crude oils.

woody An informal term used by some analysts in kerogen typing to describe *kerogen derived from the lignified tissues of land plants, i.e. wood. It generates gas at the appropriate maturity levels, and is synonymous with *vitrinite and partly with *humic. It equates chemically with *Type III kerogen.

X A C_{30} pentacyclic *triterpane of ambiguous structure which indicates derivation from a higher plant source. The compound occurs in the *saturate fraction of oils and *extracts and can be detected by *gas chromatography–mass spectrometry at, *m/e* 191, where it elutes after the C_{29} norhopane. It may have a methyl homologue, which elutes after the C_{30} *hopane. The relative abundance of the compound appears to increase with increasing maturity. *Reference*: Philp, R.P. and Gilbert, T.D. (1986). Biomarker distributions in Australian oils predominantly derived from terrigenous source materials. *Org. Geochem.*, **10**, 73–84.

Y An unidentified C_{27} *terpane which elutes close to C_{27} trisnorhopane (T_m), during *gas chromatography–mass spectrometry analysis of the *saturate fraction of oils and *extracts, at *m/e* 191. Its presence indicates derivation from a higher plant source. The compound may have a methyl homologue which elutes close to the C_{28} hopane (*bisnorhopane). The relative abundance of the compound appears to increase with increasing maturity. *Reference*: Philp, R.P. and Gilbert, T.D. (1986). Biomarker distributions in Australian oils predominantly derived from terrigenous source materials. *Org. Geochem.*, **10**, 73–84.

***Z* and *E* isomerism** *See* geometrical isomerism.

zeolite A naturally occurring, crystalline, aluminosilicate mineral. Structurally it consists of a three dimensional network of tetrahedra linked by *oxygen atoms. This framework of tetrahedra forms channels which can accommodate molecules of diameters less than the size of the channels. Different zeolites have different channel sizes. Zeolites can be made artificially to give a large range of channel sizes, e.g. Mordenite, 8Å. They are used as *molecular sieves, ion exchange resins, and drying agents. *See also*: molecular sieve.